T0234801

THE BEDFORD SERIES IN HISTORY AND CULTURE

Brown v. Board of Education

A Brief History with Documents

Related Titles in
THE BEDFORD SERIES IN HISTORY AND CULTURE
Advisory Editors: Natalie Zemon Davis, Princeton University
Ernest R. May, Harvard University

THE BEDFORD SERIES IN HISTORY AND CULTURE

Brown v. Board of Education

A Brief History with Documents

Edited with an Introduction by

Waldo E. Martin Jr.

University of California, Berkeley

Palgrave Macmillan

For Palgrave Macmillan

History Editor: Katherine E. Kurzman
Developmental Editor: Charisse M. Kiino
Production Editor: Tony Perriello
Marketing Manager: Charles Cavaliere
Production Assistant: Deborah Baker
Copyeditor: Barbara G. Flanagan
Text Design: Claire Seng-Niemoeller
Indexer: Steve Csipke
Cover Design: Richard Emery Design, Inc.
Cover Art: Portrait of Nettie Hunt and daughter, Nikie, 3½, on the steps of the Supreme Court, 1954. Washington, D.C. Courtesy of CORBIS/UPI.
Composition: ComCom
Printing and Binding: R. R. Donnelly and Sons

President: Charles H. Christensen
Editorial Director: Joan E. Feinberg
Director of Editing, Design, and Production: Marcia Cohen
Managing Editor: Elizabeth M. Schaaf

Library of Congress Catalog Card Number: 97–74964

Manufactured in the United States of America.

2   1   0   9   8
f   e   d   c   b   a

For information, write: Bedford/St. Martin's, 75 Arlington Street, Boston, MA 02116
(617-426-7440)

ISBN 978-0-312-11152-6 (paperback)
ISBN 978-1-349-61227-7   ISBN 978-1-137-07126-2 (eBook)
DOI 10.1007/978-1-137-07126-2

Acknowledgments

Petition on Behalf of Black Inclusion in the Boston Common Schools (October 17, 1787). From *A Documentary History of the Negro People in the United States,* volume 1, Herbert Aptheker, editor. Copyright © 1961. Published by arrangement with Carol Publishing Group. A Citadel Press Book.

Maria W. Stewart, *A Black Teacher's Travail* (1850s). Reprinted with the permission of Indiana University Press. "Sufferings During the War," from Marilyn Richardson, ed. *Maria W. Stewart: America's First Black Woman Political Writer—Essays and Speeches* (Bloomington: Indiana University Press, 1987).

Fugitive Slave Poster. Prints and Photographs Division, Library of Congress.

Acknowledgments and copyrights are continued at the back of this book on page 245–46, which constitutes an extension of the copyright page. It is a violation of the law to reproduce these selections by any means whatsoever without the written permission of the copyright holder.

Foreword

The Bedford Series in History and Culture is designed so that readers can study the past as historians do.

The historian's first task is finding the evidence. Documents, letters, memoirs, interviews, pictures, movies, novels, or poems can provide facts and clues. Then the historian questions and compares the sources. There is more to do than in a courtroom, for hearsay evidence is welcome, and the historian is usually looking for answers beyond act and motive. Different views of an event may be as important as a single verdict. How a story is told may yield as much information as what it says.

Along the way the historian seeks help from other historians and perhaps from specialists in other disciplines. Finally, it is time to write, to decide on an interpretation and how to arrange the evidence for readers.

Each book in this series contains an important historical document or group of documents, each document a witness from the past and open to interpretation in different ways. The documents are combined with some element of historical narrative—an introduction or a biographical essay, for example—that provides students with an analysis of the primary source material and important background information about the world in which it was produced.

Each book in the series focuses on a specific topic within a specific historical period. Each provides a basis for lively thought and discussion about several aspects of the topic and the historian's role. Each is short enough (and inexpensive enough) to be a reasonable one-week assignment in a college course. Whether as classroom or personal reading, each book in the series provides firsthand experience of the challenge—and fun—of discovering, recreating, and interpreting the past.

Natalie Zemon Davis
Ernest R. May

Preface

The importance of the 1954 *Brown* decision cannot be overstated. It is often characterized as the most important Supreme Court ruling in our history. In a singularly bold stroke, *Brown* reversed a line of legal support for Jim Crow—the invidious doctrine of separate but equal public accommodations and institutions for blacks and whites. The institutionalization of "separate but equal" achieved its most extensive expression in the American South during the first two-thirds of the twentieth century, but its essence was practiced concurrently nationwide.

Far too often *Brown* is the subject of ritual praise in the absence of knowledge of its rich and revealing history: its origins, development, meanings, and consequences. This volume offers a twofold historical examination meant to address this deficiency. It consists of an interaction between a critical and analytic narrative, on one hand, and essential documentary evidence charting a trajectory for *Brown,* on the other. The historical framework provided here melds interpretation and evidence. As a result, this historical exploration is meant to provoke discussion and debate.

This text presents *Brown* as a stunning success within the ongoing African American freedom struggle: the triumphant culmination of an intense legal and social struggle against Jim Crow specifically and antiblack oppression generally. Throughout, the complexity of that success is emphasized: its costs as well as its benefits. The actual decision handed down by Chief Justice Earl Warren on May 17, 1954, declaring racially segregated schools inherently unequal and therefore illegal was the result of an extraordinary series of concerted efforts by the legal staff of the National Association for the Advancement of Colored People.

This volume treats *Brown* as both a historical watershed and a powerful cultural symbol, or metaphor. This approach sheds light on the parallel and related emergence of both post–World War II American hegemony and the "movement" (civil rights and Black Power). Indeed *Brown*'s energizing impact on the midcentury African American liberation insurgency proved absolutely fundamental. Although the black freedom struggle has deep historical antecedents—ultimately back to the point of enslavement—the starting point in this volume is the mid-nineteenth century in *Roberts v. City of Boston* (1849), a significant case in the legal history of racial segregation in schools. In addition to the critical legal documents, this volume also includes relevant social and cultural documents. Together, these two kinds of materials are mutually revealing and deepen our understanding of the historical context surrounding the decision's evolution over time.

Three interrelated themes unite the threads of this project. First, this text considers a continuing American dilemma: the ongoing struggle of the American nation to fully incorporate its African American citizens. Second, it sheds light on the enduring nature of the African American freedom struggle itself. Indeed, as illustrated throughout, one must understand the symbiotic relationship between these two struggles. Third, it highlights the historical complexity and historical significance of both of these struggles. The legal documents frame the historical, constitutional, and juridical debates surrounding the theory and practice of antiblack discrimination.

Brown reflects a significant cultural shift as well as a key historical turning point. It is useful to consider what *Brown* means as an event, as a symbol, and as a key marker in the ongoing black liberation struggle. It is obviously far more than a legal decision outlawing racial segregation in schools. It signifies a critical juncture in the postwar refashioning of the United States into a nation truer to its self-image as one nation out of many. *Brown* is a signal and a most revealing moment in the continuing historical and cultural making of the American people, notably the Americanization of blacks.

ACKNOWLEDGMENTS

This volume benefited significantly from the assistance of numerous individuals and institutions. The readers of the initial draft for the press—Mark Tushnet, Genna Rae McNeil, Lynn Dumenil, Thelma Foote, and two anonymous scholars—offered expert and helpful comments. A scholarly coterie of colleagues and friends—Patricia Sullivan, Shelia Martin, Michelle Deardorff, Rickey Hill, M. Rose Gladney, Thandekile R. M. Mvusi, Timothy Huebner, Leslie McLemore, Charles Vincent, Renee Romano, and James SoRelle—offered useful responses to the introduction. A special graduate student seminar organized by Scott Tang at

Berkeley and led by Robert Avila, Karen Leong, Tamson Chumbley, Ronald Lopez, Timothy Lynch, Donna Murch, and Diana Williams likewise provided thoughtful reactions to the introduction. Elizabeth Abel of Berkeley's English Department graciously shared her knowledge of Jim Crow photography and vital copies of prints. William Moore of the College of Charleston's Political Science Department generously provided a key document at the final hour.

Of the various libraries and services I consulted at the University of California at Berkeley in the course of the preparation of this volume, several merit special mention: Charles F. Doe (the main library), Moffitt Undergraduate Library, Baker Document Delivery Service, Interlibrary Loan, and the Law Library. Phyllis Bischof, librarian for African and African American Collections, was extremely helpful. Stanford's Law Library, Harvard's Widener and Lamont Libraries, and Radcliffe's Schlesinger Library also provided first-rate service.

Elizabeth Gessel was my research assistant for the bulk of this project. Her smart and wide-ranging assistance was invaluable. Toward the end, Kirpal Johnson performed the same role with equal aplomb. At the very end, Peter Lau helped out, too. Gail Phillips, Jimmy Brown, and Adriane Thrash in Berkeley's History Department provided critical word processing help.

At Bedford Books, several persons supported this project in its various stages and deserve thanks: Charles Christensen, Joan E. Feinberg, Elizabeth M. Schaaf, Katherine Kurzman, Sabra Scribner, Niels Aaboe, Tony Perriello, Terry Govan, Richard Emony, and Barbara Flanagan. I am especially indebted to the discerning editorial eye of Charisse Kiino.

I must thank my daughters, Jetta Grace Martin and Coral Rose Martin, for their faith in the project and their understanding that the time it took was worthwhile. Finally, Catherine L. Macklin, my wife, is my best friend, my best critic, and soul inspiration.

Waldo E. Martin Jr.

Contents

Brown v. Board
of Education

A Brief History with Documents

Introduction:
Shades of *Brown:* Black Freedom,
White Supremacy, and the Law

Arguably the most important Supreme Court ruling in United States history, the *Brown* decision in 1954 not only overturned the doctrine of separate but equal schools as unconstitutional, but it also put other forms of antiblack discrimination on the road to extinction. The unanimous decision reversed the Court's 1896 decision in *Plessy v. Ferguson,* which had upheld the concept and practice of state-endorsed racial discrimination — Jim Crow — the chimera of separate but equal public accommodations and institutions for blacks and whites. The *Brown* decision was the culmination of countless interrelated collective and personal battles waged by blacks and of a series of legal efforts by the National Association for the Advancement of Colored People (NAACP) from the early days of its existence in the 1910s and 1920s.

Indeed, the legal cases that have influenced the status of African Americans historically have come out of the day-to-day struggles of regular people, such as those in Clarendon County, South Carolina, whose fight for better black schools in the late 1940s became one of five cases to be ultimately joined as *Brown v. Board of Education.* The segregated schools for blacks in Clarendon County at the time were a disgrace, clearly worse than most all-black schools in the South. Black life in the county was extremely hard. In his definitive work *Simple Justice: The History of* Brown v. Board of Education *and Black America's Struggle for Equality,* Richard Kluger notes that "if you had set out to find the place in America in . . . 1947 where life among black folk had changed the least since the end of slavery, Clarendon County is where you might have come." In 1950, more than two-thirds of the county's black households earned less than $1,000. The county maintained twelve schools for whites and sixty-one for blacks. Over half of the black schools were shanties with a teacher or two and a student body ranging widely in age and educational level.

In 1950, the total value of the black schools was $194,575; that of the white schools was $673,850. For the 1949–50 school term, the county school board spent $43 per black child, $179 per white child. Black teachers earned two-thirds less than their white counterparts.[1]

In 1947, black parents, led by Reverend J. A. DeLaine and Hugh Pearson, a local farmer, began pressing the county to provide buses for black students as it already did for white students. By the following year, with the help of local black lawyer Harold W. Boulware and the local and national branches of the NAACP, the struggle had escalated dramatically, with a lawsuit in federal court. Argued by Thurgood Marshall, the head of the NAACP legal defense team, the lawsuit demanded that the county go beyond equalizing its black and white schools and fully integrate its public school system. The plaintiffs in the suit, Liza and Harry Briggs, lost their jobs as maid and service station attendant, respectively; despite other instances of white repression of local blacks, the legal battle went forward.[2]

Briggs v. Elliott soon joined four similar cases argued by the NAACP's legal team before the Supreme Court: *Brown v. Board of Education of Topeka, Kansas; Davis v. County School Board of Prince Edward County* (Virginia); *Belton v. Gebhart* (Delaware); and *Bolling v. Sharpe* (District of Columbia). In late 1952, the Court consolidated and first heard these cases under the rubric of *Brown*. Public school segregation, according to the NAACP's legal brief, was a violation of the Fourteenth Amendment's equal protection clause. An integral element of the effort to make blacks part of the nation during the Reconstruction period (1863–77), this 1867 amendment clearly defined U.S. citizenship to encompass all blacks. Furthermore, it stated that all citizens were equal under the law. Consequently, the NAACP lawyers argued, the blatantly unequal racially segregated schools were unconstitutional and had to be integrated. In each case, local lawyers in conjunction with NAACP lawyers sought the immediate end of Jim Crow schools as intrinsically separate and unequal. The lawyers also argued that state-sanctioned segregation stamped blacks with a stigma of inferiority that undermined their self-esteem. In effect, the aim in *Brown* — the total and unconditional abolition of Jim Crow schools — represented a critical move in the black freedom struggle.

HISTORICAL BACKDROP: THE CONSTITUTION, THE LAW, AND FIGHTING JIM CROW

The continuing African American freedom struggle has produced simultaneous, often overlapping, battles on many fronts: political, economic, social, cultural. The nineteenth-century war against slavery featured moral as well as political and economic campaigns in the North. This effort required that slaves and free blacks in the South struggle as best they could until conditions were ripe for seizing freedom for the enslaved, as they were in the Civil War. Abolitionism—various northern-based movements to emancipate the slaves—was a divisive national issue that alienated most whites, especially the proslavery forces in the North and South. Like the war against Jim Crow, abolitionism was a war waged by a dissenting black minority and its stalwart white supporters: a multifaceted and far-reaching struggle.

The history of the *Brown* decision constitutes an integral flank in the war against Jim Crow. While the story of *Brown* can be presented in various ways, the chronological emphasis here is twofold. A long-term perspective stresses the first half of the twentieth century, the *Plessy* period. A short-term perspective stresses post–World War II America, especially the late 1940s through 1955—the immediate historical context of *Brown*. To comprehend the shift from *Plessy*—the world of separate and unequal caste relations between blacks and whites—to *Brown*—the world of legal equality—more historical background is imperative.

In the antebellum period, the black slaves' lack of legal rights, combined with the severely circumscribed legal rights of free blacks in both the North and South, made the constitutional status of both groups precarious at best. The Supreme Court decision in *Scott v. Sandford* (1857) codified the "quasi-free" and degraded legal status of free blacks with its argument that blacks, free and slave, lacked legal standing in U.S. courts.[3] Not surprisingly, therefore, well before the Civil War, emancipation, and Reconstruction, antiblack prejudice and discrimination was rife. It is clear that the racism of the Jim Crow era had important antecedents in earlier ideologies and structures of white supremacy.

These structures included severe job discrimination; political exclusion including disfranchisement; civic disabilities, such as exclusion from juries and prohibitions against black testimony in courts (typically in cases involving black testimony against whites); social ostracism and residential segregation; exclusion from the public domain of schools, churches, hotels, restaurants, pubs, halls, and conveyances; and antiblack terrorism. Proslavery ideology espoused the "civilizing" influence of bondage for Africans, and polygenetic thought argued for multiple and separate human species, with Africans representing an inferior order of creation unrelated to that of whites. In the arena of popular culture, blackface minstrelsy, with its racist and demeaning caricatures of blacks, was the most popular form of mass entertainment in the nineteenth century.[4]

Emancipation and Reconstruction imperfectly incorporated African Americans into post–Civil War America. These developments encompassed upgrading the constitutional status of all blacks, leading to passage of the very important Reconstruction amendments to the Constitution: the Fourteenth Amendment (1867) defining citizenship and providing equal protection under the law, the Thirteenth Amendment (1865) outlawing slavery, and the Fifteenth Amendment (1870) removing race as a qualification for voting. This effort to rectify the disparity between the racist oppression of blacks and the American creed of freedom, justice, and equality for all was important, but it was part of a deeply flawed Reconstruction experiment.

The most crucial defect was the failure to provide emancipated blacks with an economic foothold in society — land of their own and the economic wherewithal to make a decent living. During the Reconstruction period, new forms of economic dependency, such as sharecropping and tenantry, developed, often leading to debt peonage, a new kind of "economic slavery."[5] In fact, slavery's stigma — the mark of black racial inferiority — persisted, reinforcing the systemic economic exploitation of black labor. The struggle against Jim Crow was fundamentally as much about economic justice as about political and civil rights.

As Reconstruction gave way to the restoration of white power in the South — a process described as "Redemption" — the brief yet important moment of black political participation in the nineteenth-century South began to close. Racist intimidation and violence buttressed the restoration of white rule in the South. By the turn of the century, the exclusion of blacks from the political process was pervasive, through devices like poll taxes and the "white primary." The latter made the southern state Democratic parties into private clubs legally able to exclude blacks. This racist counterinsurgency effectively disfranchised southern blacks.[6] As a result, political struggle through means like lobbying the president and Congress, spreading the struggle through the press and other forms of mass media, boycotts, and petitions became indispensable. Similarly, the lack of formal political influence made informal political struggle through instruments like the courts all the more necessary.

Complementing the formal political erasure of blacks in the South, the geographic and psychological exclusion basic to Jim Crow was a vital component of the turn-of-the-century resurgence of white supremacist belief and practice. In a very real sense, the spread of Jim Crow went hand in hand with the growth of an increasingly influential scientific and intellectual racism: a kind of cultural "neoslavery." From the nineteenth-century practice of measuring and comparing racial skulls to the twentieth-century fixation on IQ tests, the rage to offer "scientific" explanations for alleged racial difference has continued to this day.[7]

Concurrently, popular Western discussions of the "white man's burden" confined African Americans and peoples of color to their place in the Victorian social hierarchy: structurally and intellectually below their racial superiors, whites. There, according to this racist ideology, blacks and their racial cohorts could rise to the level of civilization commensurate with their inferior capacities. Social Darwinism—a crude late-nineteenth-century sociobiological rationalization of the status quo, notably its inequities—justified this virulent racism as consistent with the natural order: the social "survival of the fittest" and the inevitable dominance of the superior race.[8] This is the world that produced Jim Crow.

Largely as a white counterreaction to the perceived threat of post-bellum black freedom, then, white supremacy reasserted itself with a vengeance in the late nineteenth century. Jim Crow's turn-of-the-century rigidification and codification exemplified this far-reaching, often violent reassertion. The realities of Jim Crow furthered the development of separate black and white worlds, economically, socially, and culturally. The institutional infrastructure within black communities expanded to address those communities' most pressing concerns. This could perhaps best be seen in the growth of distinctive black business worlds—including black morticians, barbers and hairdressers, bankers, and insurance companies—catering almost exclusively to a black clientele.

Jim Crow also meant the brutal suppression of successful as well as assertive blacks as "uppity" and "getting out of their place." The responsibility of protecting the privilege of whiteness demanded no less. These attitudes sustained the horror of growing numbers of rapes of black women by white men and white lynchings of black men falsely accused of raping white women. Intertwined myths of black animality and black hypersexuality justified these crimes against black humanity. The turn of the century was clearly a nadir for American race relations.[9]

In spite of the terrorism of Jim Crow, the black freedom struggle persisted. In the twentieth century as before, most black activism has been local and often unheralded, notably before the emergence of the modern civil rights movement (1954–65) with its attendant national, international, and mass-media spotlights. Innumerable battles were waged not only in the courts but in city halls, county boardrooms, state and national legislatures, and various executive offices as well as sites like the streets, the job, and prison—within institutional structures and outside them, as in the civil disobedience campaigns of the civil rights movement. The various strategies — bus boycotts, sit-ins, marches — developed in response to specific historical challenges and contexts.

Of course, this increasing black politicization owed in part to many efforts, often earlier ones, including those of the Universal Negro Improvement Association, the mass black nationalist organization created by Jamaican Marcus Garvey that peaked in the early 1920s. Black political education and political organization expanded in response to the economic devastation of the Depression during the 1930s. While black allegiance shifted to the Democrats in response to the hope spawned by Franklin D. Roosevelt's New Deal, equally important was the burgeoning sense that a better world for blacks could be realized through renewed organization and agitation. The Communist Party's support for the black freedom struggle also encouraged growing black optimism regarding positive social change.

The black freedom insurgency grew dramatically during World War II as black protest, black mobilization within organizations like the NAACP, and the democratic and egalitarian rhetoric of the official war effort emboldened African Americans. During the war blacks created the Double Victory Campaign: victory over domestic racism — evil at home — as well as Nazism, fascism, colonialism, and international racism — evil abroad. The most important component of the ideological struggle, however, was domestic: the battle to realize the American creed.

A key aspect of this shifting context was the heightening awareness among increasing numbers of blacks of the international ramifications of their domestic struggle. For many, commitment to democracy in the United States demanded alignment with progressive struggles for self-determination, especially in the Third World. For more and more black freedom fighters as well as American politicians, postwar American apartheid was no longer domestically or geopolitically viable. The growing cold war between the Soviet Union and the United States highlighted the blatant contradiction between the American creed and the reality of America's treatment of its black citizens. As the putative leader of the "free world," the United States had to get its domestic house in order. World leadership in an international community made up more and more of nonaligned Third World nations — nations created principally by peoples of color — rendered Jim Crow unacceptable.

The declining legitimacy of racism in the United States stemmed from a mix of forces, both domestic and international, which came together at midcentury. The *Brown* decision crystallized and reflected the pivotal move toward racial equality: the rejection of white supremacy as signified in *Plessy*. In this way, the law assumed a critical role in the political and intellectual as well as constitutional struggles against racism.

Elite venues—lawyers' offices, courtrooms, judges' chambers—are principal sites of formal legal struggles. Nevertheless, the legal struggle—battles waged from the top down—does not happen in isolation but exists in dynamic mutuality with the social struggle—battles waged from the bottom up. This mutuality was especially vital in *Brown* where individual cases came together as class actions seeking relief for all members of the aggrieved class, in this instance all African Americans subject to Jim Crow discrimination. A crucial aspect of the civil rights litigation pioneered by the NAACP legal defense team was the guiding assumption that the legal battles were part and parcel of the collective struggle of African Americans.

THE EVOLUTION OF THE NAACP LEGAL CAMPAIGN AGAINST JIM CROW

"The problem of the twentieth century," African American scholar and leader W. E. B. Du Bois perceptively noted in 1903, "is the problem of the color line." This often quoted prophecy speaks directly to the uncharted road leading to *Brown:* how best to alleviate the "color line," a primary manifestation of Jim Crow. At this time, Du Bois increasingly favored public agitation, particularly political organization and political action. Booker T. Washington, the most famous and influential black leader from 1895 to 1915, publicly advanced an accommodationist strategy emphasizing black self-help and the cultivation of goodwill between the races, not agitation. These two positions reflected a continuing debate on how best to advance the interests of African Americans.[10]

In spite of Washington's accommodationist public persona, behind the scenes he was actively engaged in lawsuits challenging Jim Crow. Du Bois, however, took the more publicly activist route and along with prominent blacks such as Ida B. Wells and William Monroe Trotter and whites such as Mary White Ovington and Oswald Garrison Villard launched the interracial National Association for the Advancement of Colored People (NAACP) in 1909. Founded in part as a response to a series of antiblack race riots, most notably the 1908 Springfield, Illinois racial conflagration, the NAACP soon emerged as the leading black civil rights organization.[11]

The NAACP pursued several lines of attack in its assault on the "color line." Lobbying for favorable legislative, judicial, and executive action; waging a publicity war through the media, most effectively in the *Crisis* magazine, initially edited by Du Bois; and working with grassroots chapters on specific issues of local concern such as discriminatory ordinances, the organization endeavored to advance a black civil rights agenda. Intensely fought battles against antiblack discrimination in jobs, housing, voting, public accommodations, and education demanded functional knowledge, savvy, and flexibility. Given its limited resources and the awesome power of the racist status quo, the NAACP favored significant yet workable battlegrounds where its members could realistically achieve the upper hand. Victories with far-reaching impact were thus highly desirable.[12]

From the beginning, litigation proved to be a particularly important and effective tool in the organization's armament. The legal struggle against segregated schools in mid-nineteenth-century Boston and Jim Crow railway cars at the turn of the century clearly presaged the NAACP legal campaigns. In the Boston school integration (1849), *Dred Scott* (1857), and *Plessy* (1896) cases, the decisions went against the individual black claimants and the collective aspirations of blacks. Nevertheless, hope remained that the rule of law would eventually be squared with constitutional claims for full black equality, especially following the enactment of the Fourteenth Amendment.

The legal endorsement of equality in *Brown* was a capstone to an extraordinary series of battles against de jure (legal) and de facto (actual) Jim Crow. The Fourteenth Amendment's guarantee of equal protection under the law epitomized the legal tradition undergirding *Brown*. Early American legal tradition (1787–1830) was built on English common law and emphasized freedom, equality, and justice for all citizens as framed in the Constitution (1787) and Bill of Rights (1791). With its powerful Enlightenment grounding, this compelling vision of constitutional law stressed reason, order, and progress as inseparable from freedom, equality, and justice. The United States ideologically embraced a republican form of government that deepened the young nation's commitment to these tenets.

This libertarian, or pro-freedom, reading of the Constitution and the law is fundamentally antithetical to the slavery and racism the nation's founders embraced. In fact, the founding patriarchs countenanced freedom for whites fully predicated upon black slavery and black debasement. This haunting paradox has decisively shaped the American nation since its founding.[13] However, the libertarian view of the Declaration of Independence (1776) and the Constitution, along with the radical egalitarianism of the former, provided indispensable ideological bases for the black freedom struggle from the beginning until now. The documents collected here capture the historic tension between law and social practice rooted in white supremacy on the one hand and human equality on the other. As a result, they can be analyzed from at least two contrasting points of view—that of the white supremacist status quo and that of the black freedom struggle.

This work is also a discussion about the nature of the law, especially what is meant by the rule of law. To what extent is law organic and timeless? To what extent is it created and manipulated? What is the relationship between the rule of law and justice or fairness? How do issues of race impinge upon the law? What is the relationship between legal developments and historical context? The texts here provide a forum for the consideration of these kinds of questions. Through a critical history of *Brown*, this collection throws much-needed light on interrelated social, cultural, political, and economic as well as legal developments which gave rise to this epochal decision.

A deep-seated belief in the prospects for advancing black civil rights through the legal system earmarked the highly influential career of Charles Hamilton Houston, who was most responsible for charting the various legal paths that led to *Brown*. In 1983, Judge A. Leon Higginbotham Jr. wrote that "Houston was the chief engineer and the first major architect of the twentieth-century civil rights legal scene." He "almost single-handedly . . . organized and led the legal battalion in the critical early battles seeking equality for black Americans."[14]

Harvard-trained and the first black elected to *Harvard Law Review*, Houston left a private practice he shared with his father in Washington, D.C., to become dean of Howard University's law school (1924–35). Houston put that institution on sound academic footing, making changes that led to the school's full accreditation and greatly enhanced its prestige. Among his numerous accomplishments, several bear mentioning. First, he trained many talented black lawyers at a time when there were precious few. Besides Thurgood Marshall, who would be instrumental in *Brown* and later would be a Supreme Court justice, Houston taught a number of prominent attorneys who would distinguish themselves in civil rights litigation, including Edward P. Lovett, James G. Tyson, Oliver W. Hill, Coyness L. Ennix, and Leslie S. Perry.[15] Second, he pioneered in two fields of legal study and practice: civil rights law and public interest law. Third, he engaged in a whirlwind of civil liberties, civil rights, and antidiscrimination activities, beyond his university duties, including the defense in the highly publicized Scottsboro case in which nine young Alabama blacks were accused of raping two white women on a freight train.[16] By 1935, Houston had emerged as the most influential black lawyer in the United States.

In light of that status, it is not surprising that when the NAACP sought a new special counsel in 1934, Houston was chosen. Having taught law and litigated a variety of cases, he was now charged with the responsibility of directing the litigation activities of the most important black civil rights organization in the country. Houston stressed that the law was a potentially useful means to promote social change, especially in the context of a complicated social struggle. Why the judicial system? As historian Genna Rae McNeil has noted: "With little power to compel congressional or presidential concessions and with virulent racism ever a possible consequence of direct action, blacks were in a better position to seek redress through the courts."[17]

Limitations of the judiciary tempered the optimism of Houston and other civil rights lawyers, however. They fully understood that historically the law had been principally a conservative and at times reactionary force. They were deeply aware of what historian Mary Frances Berry has aptly referred to as "constitutionally sanctioned violence against blacks and violent suppression of black resistance—the outgrowth of a government policy based on essentially racist, not legal, concerns—throughout the American experience." In other words, whites used "the Constitution in such a way as to make law the instrument for maintaining a racist status quo."[18]

Houston's view of the lawyer as a social engineer owed heavily to his fervent commitment to the black freedom struggle and his belief in the integral relationship between that social insurgency and legal activism. The black lawyer, according to Houston, had to envision and to practice law as a mechanism for progressive social change. A modern "race man," he fully understood that black lawyers had a special mission to fight their own people's battles. They could not depend on the white-dominated legal guild—given its historic support for white supremacy—to fight for black rights. It was imperative, according to Houston, that the black lawyer embrace

> . . . the social service he can render the race as an interpreter and proponent of its rights and aspirations. . . . Experience has proved that the average white lawyer, especially in the south, cannot be relied upon to wage an uncompromising fight for equal rights for Negroes. He has too many conflicting interests, and usually himself profits as an individual by that very exploitation of the Negro, which, as a lawyer he would be called upon to attack and destroy.[19]

Houston's adherence to the legal realism of his Harvard mentors Roscoe Pound and Felix Frankfurter provided a powerful intellectual framework for his activist legal philosophy. According to the sociological jurisprudence of legal realism, law served particular social interests; it reflected the biases and predilections of those who made and interpreted it. Legal realism, a view first fully enunciated by the eminent Supreme Court Justice Oliver Wendell Holmes Jr. (1841–1935) earlier in his legal scholarship, rejected the dominant and traditional view of the law as a set of formal rules deducible from abstract concepts like justice. Whereas legal tradition inspired judicial restraint, legal realism—especially as articulated by Houston—inspired judicial activism.[20]

Houston's legal realism complemented and energized his view that black lawyers had to be social engineers. These interlocking philosophies had an enormous impact on the Howard law curriculum, the lawyers he trained and influenced, the legal philosophy of the NAACP, and the activism of those engaged in black rights litigation. In effect, social engineering through law meant the use of the law itself wherever possible to solve the problems confronting blacks.[21]

Houston's legal reasoning authenticated the use of sociological evidence when arguing against segregation. A key example was the use of social psychological data to argue the harmful effects of racism on whites and blacks (see p. 142 in this book). Persuasive challenges to public and social policies which braced Jim Crow became an important objective of this brand of sociological jurisprudence. These kinds of legal arguments also gave focus and shape to the burgeoning field of civil rights law. As legal scholar Mark Tushnet notes, "The constitutional argument against segregation could be keyed to facts and policy." Tushnet concludes that "the sociological argument was Realist to the core. Law, even constitutional law, was social policy. Social policy had to be understood as it actually operated."[22]

Houston, Marshall, and the many other lawyers and activists engaged in the war against segregation understood that victory could not be won solely in the courts, but only through a broad-based attack. Marshall, who succeeded Houston as NAACP general counsel in 1939, relied on his mentor's counsel until Houston's untimely death in 1950. He continued to elaborate on his mentor's social engineering framework throughout his distinguished legal career.[23] Both men envisioned litigation as a tool to educate and politicize the public, white and black, about the black freedom struggle and the role of the judiciary in advancing that cause. The NAACP's legal campaign, therefore, was not a series of uncoordinated court battles, but an integral part of a much broader philosophy of social insurgency. In part, this legal campaign functioned as a mechanism to publicize the work of the NAACP and in turn to recruit members for the organization and the black freedom struggle generally.[24]

Believing that carefully executed litigation could contribute to local grassroots activism and the development of a mass movement, the NAACP and its legal staff supported local legal struggles. NAACP lawyers worked hand in glove with local lawyers, whether the issue was black political exclusion, disparities between white and black teachers' salaries, a black falsely accused or convicted of a crime, or some other miscarriage of racial justice. The NAACP also mounted a vigorous legal and educational campaign against the most virulent forms of legal racism, such as the highly visible terrorism of state-sanctioned white rule through mob action and lynch law. In far too many instances in the first half of the twentieth century, a black accused of a crime—especially a black man accused of raping a white woman—was murdered publicly by angry white lynch mobs with no concern for niceties like court trials or convictions. Although the NAACP had waged an unrelenting and highly public battle against white lynch law since 1919, the group—like others struggling against this heinous injustice—was unable to persuade the federal government to pass an antilynching law. Southern white opposition, notably in the Congress, effectively blocked all such efforts.

Battered but undaunted, the NAACP went forward. The seemingly impregnable state-sanctioned world of Jim Crow fueled extensive debate within the organization around what tactics to use to dismantle institutionalized racism. Two related debates in the 1930s illuminate the nature and impact of this spirited discourse: (1) the kind of legal strategy to pursue and (2) more broadly considered, legalism versus alternative strategies.

The first debate was over whether the NAACP lawyers should attack the entire edifice of Jim Crow forthrightly by seeking a ruling nullifying *Plessy*—a direct attack strategy—or, whether they should work incrementally, building a series of legal victories that paved the way for the eventual dismantling of *Plessy*—a developmental strategy. A principal goal of the developmental strategy was to force the South to equalize its separate black and white worlds through litigation by making Jim Crow fiscally and politically unworkable. Given the relative poverty of the South and the declining respectability of Jim Crow, equalization would undermine American apartheid.

Nathan Margold, Houston's predecessor as head of the NAACP legal team, had pushed for the direct attack strategy. Like Houston, Margold was a protégé of Felix Frankfurter and was committed to both legal realism and judicial activism. In 1931, a year after his hiring, Margold issued a bold report strategically arguing for a direct attack on segregation, leaving open the issue of equalization. A frontal assault would cut immediately to the heart of the issue — cogent legal demonstration of the fundamental wrong of state-sanctioned racial segregation — and would require an immediate end to Jim Crow. Margold preferred that the issue of equalization be treated as a related but subordinate concern.

Margold maintained that a direct attack was preferable as it required fewer suits and the NAACP's legal staff could devote its attention to precedent-setting cases. Similarly, this approach avoided litigating overlapping suits at the state and local levels and thus the often confusing and conflicting welter of federal, state, and local statutes. Also, as the Margold report explained, a direct attack was a better use of the NAACP's limited fiscal resources and its small legal staff.[25]

Houston and Marshall after him firmly believed that the Margold report put forth a position which the NAACP and the larger black freedom struggle should support in theory; however, in reality, they realized that the times were inauspicious for such an aggressive strategy. In the Depression years, economic hardships intensified among blacks and spread among whites. Economic turmoil further exacerbated racial tensions and did not provide the most supportive setting to battle Jim Crow. In addition, the NAACP lacked sufficient mass black support and progressive white support on the one hand and the necessary strategic support within the legal establishment on the other. In the 1930s in particular, many blacks still had to be convinced that a legal assault against Jim Crow was viable. Otherwise, local blacks facing the extraordinary pressures brought to bear against those who filed anti–Jim Crow suits might not have the uncommon courage and the black community support necessary to proceed. Another important impediment which had to be overcome was the widespread lack of trained black lawyers.[26] Therefore, Houston decided to employ a more moderate strategy of equalization as a way to build support for a direct attack later. Cultivating a network of popular and professional support became a vital tactical goal.

Houston chose to focus the legal assault on education because of its centrality to advancement and fulfillment within American culture. As such, the blatant denial of equal educational opportunities to black youth touched a powerful nerve in the American psyche. The terrible realities of segregated education in the South offered compelling evidence of gross racial disparities in facilities, budgets, and salaries. Also, Houston contended, "discrimination in education is symbolic of all the more drastic discriminations,"[27] such as lynch law. Furthermore, Jim Crow education represented the deeply ingrained stigma of innate black racial inferiority.

Houston's strategy featured three related aspects. Desegregation of public graduate and professional schools was one. Here the battle was fought at a less contentious level than that of elementary and secondary schools. Equalization of white and black teachers' salaries was the next aspect. The NAACP legal team achieved a number of important victories in salary cases. As a result, many southern school boards masked salary differentials through the use of so-called merit criteria, and the cases became much harder to argue. It was not until the late 1940s that the next level of the legal plan — equalization of elementary and secondary school facilities — became feasible. Until then, overcoming the local and tactical obstacles hindering these cases proved too difficult.[28]

Another challenge was finding and sustaining the morale of litigants whose character and resources would have to withstand intense public scrutiny and white reprisals — typically economic, sometimes physical and violent. The prolongation of many cases caused litigants to lose enthusiasm and even drop out. Racist southern school districts used various legal strategies to tie up the proceedings and to exhaust black litigants financially and emotionally. Often these districts admitted to the disparities in their educational offerings but exaggerated or lied about efforts under way to ameliorate them. The defense used this tactic in the South Carolina district court case of *Briggs v. Elliott* (see p. 126).

Other obstacles faced the legal team. First, the fact that the states and local school districts themselves were primarily responsible for public school education policy and funding inhibited litigation at the federal level. Second, with the awesome weight of tradition and social custom, *Plessy* was the precedent upon which pro–Jim Crow rulings rested. Third, it followed that courts did not consider state-sanctioned Jim Crow to violate the Fourteenth Amendment right of blacks to equal protection under the law and therefore left Jim Crow intact. Fourth, the defendants and courts alike variously ignored, trivialized, masked, neutralized, explained away, and accepted the pervasive reality of separate and unequal. All of these tactics naturalized Jim Crow as fundamental to a "higher law" of white supremacy, or integral to the organic order of society. According to Mary Frances Berry, the controlling factor in legal decisions was the ubiquity of constitutional racism. Ultimately, as Derrick Bell maintains, the law functioned to sustain white supremacy.[29]

The NAACP's Legal Strategy Challenged

It is not surprising, then, that searching questions were raised about the NAACP's growing commitment to legalism as a primary strategy: the group's second pivotal 1930s controversy. Many committed to the black freedom struggle called for greater emphasis on economic issues because of the Depression's ravaging effects. As one would expect, economic critique was widespread: it could easily be found on the street, in colleges and universities, and among radicals and progressives. Bluesman Carl Martin observed:

> Everybody's crying: "Let's have a New Deal,"
> 'Cause I've got to make a living,
> If I have to rob and steal.[30]

At the same time, economist Abram Harris and political scientist and future United Nations stalwart Ralph Bunche, both young professors at Howard University, called for interracial labor unity and an understanding of the centrality of economics, or material forces, to the historic oppression of blacks. They maintained that the oppression of blacks was not merely a problem of race but was a question of class as well. Broadly speaking, the struggle had to be one of ameliorating capitalism's most flagrant abuses. Far more oppositional, albeit less influential, voices like black Alabama Communist Party activist Hosea Hudson found capitalism itself to be the problem, socialist revolution the solution.[31]

The venerable W. E. B. Du Bois was the most provocative and powerful voice questioning the NAACP's focus in the 1930s. His perceptive critique cut two ways. First, harking back to the ideas of Booker T. Washington at the turn of the century, Du Bois now wanted the fiercely interracialist and integrationist NAACP to promote black economic development—and in turn black elevation—through aggressive support of a separate black economic world. Du Bois's Marxist-socialist-inspired critique of capitalism, calling for greater workers' control over the economy, spoke more and more of the necessity for black networks like consumer cooperatives. This message did not sit well with the intensely pro-capitalist NAACP.

Du Bois and others emphasized that legalism had to be prefaced by the redistribution of wealth and across-the-board leveling of power and influence. Reliance on legalism as a remedy for the problems confronting black Americans signaled a reformist agenda at best, they felt, certainly not a revolutionary one. After leaving the association in 1934, once the ideological rift became irreparable, Du Bois continued to offer an increasingly militant socialist and internationalist approach.[32] The "road to *Brown,*" however, was clearly being plotted through capitalism, not socialism.

Du Bois's call for black economic nationalism vividly exposed the tensions between voluntary and imposed segregation, between separatism and integrationism, between black nationalism and American nationalism. Seeking to get beyond these tensions, he stressed that blacks had to strengthen the institutional infrastructure and social fabric of their own communities. The critical issue was to forge more effective forms of collective organization and action aimed at intraracial uplift. In this vision, integration assumed a decidedly secondary, even ancillary, position. He emphasized the importance of black institutions and black culture in structuring and propelling the black freedom struggle and in nurturing the black psyche. The thrust of NAACP politics, from this point of view, increasingly now collided with rather than meshed with black needs and aspirations.

The historical and rhetorical development of *Brown* reflected a profound discomfort with racial separatism. Essential to the social-scientific discourse behind *Brown* was the argument that racial segregation, even voluntary segregation, was responsible for the psychological damage and sociocultural pathology among blacks. Du Bois clearly perceived that this negative characterization of a distinctive black life and culture as well as of blacks as victims was one-sided and misleading. This potentially baneful argument, increasingly vital to the NAACP's liberal indictment of Jim Crow, failed to make the crucial distinction between what Du Bois saw as the benefits of voluntary segregation—autonomy and psychic health—and the harm of state-imposed segregation—dependency and dehumanization.[33] The point was not that white racism had deformed black life and culture, but rather that it had deformed the American experience.

Du Bois stressed in his 1935 discussion "Does the Negro Need Separate Schools?" (see p. 91) that the fundamental issue was equality of educational opportunity: making available to black students the best education possible, whether that be in segregated or integrated schools. He explained that

> . . . the Negro needs neither segregated nor mixed schools. What he needs is education. What he must remember is that there is no magic, in either mixed schools or in segregated schools. A mixed school with poor and unsympathetic teachers, with hostile public opinion, and no teaching of truth concerning black folk, is bad. A segregated school with ignorant placeholders, inadequate equipment, poor salaries, and wretched housing, is equally bad. Other things being equal, the mixed school is the broader, more natural basis for the education of all youth. It gives wider contacts; it inspires greater self-confidence; and sup-

presses the inferiority complex. But other things are seldom equal, and in that case, Sympathy, Knowledge, and the Truth, outweigh all that the mixed school can offer.[34]

Du Bois also reiterated that the problem was white racism, not the cruel hoax of innate black inferiority. Structurally speaking, he maintained, the crux of the issue was the symbiosis between racism and capitalism. In terms of education in particular, the problem went in two directions: racist constraints on black educational opportunity, and black as well as white devaluation of black institutions and culture. In this case, the denigration of black schools and black educators, in spite of their noteworthy achievements against all odds, was common even among blacks. The brainwashing of blacks, what historian Carter G. Woodson referred to as "The Mis-Education of the Negro," was indispensable to the propaganda of white supremacy. Du Bois countered, however:

> If the American Negro really believed in himself; if he believed that Negro teachers can educate children according to the best standards of modern training; if he believed that Negro colleges transmit and add to science, as well as or better than other colleges, then he would bend his energies, not to escaping inescapable association with his own group, but to seeing that his group had every opportunity for its best and highest development. He would insist that his teachers be decently paid; that his schools were properly housed and equipped; that his colleges be supplied with scholarship and research funds; and he would be far more interested in the efficiency of these institutions of learning, than in forcing himself into other institutions where he is not wanted.[35]

Whereas Du Bois's economic and cultural nationalism did not find favor with the NAACP leadership, Du Bois and his opponents within the NAACP did agree upon the necessity of strengthening the organization's grassroots constituencies. Ordinary black folk had to be brought into the organization; they had to be made to feel that this was *their* civil rights organization. Otherwise, a black freedom struggle guided in large measure by the NAACP stood no real chance of success. Local and state branches had to be strengthened. Black politicization during the Depression and war years, especially the latter, was the seedbed of the concurrent flowering of the NAACP's membership rolls. Growing black movement toward the Democrats, the party of FDR, most notably in the North, marked this politicization. A more important signal of this trend, especially in the South where the "lily white" Democratic Party moderated the ultimately pivotal black shift toward the Democrats, was the phenomenal expansion of the NAACP.

Historian Patricia Sullivan has shrewdly observed:

Black identification with the party of Roosevelt and the revival of the NAACP were primary mediating forces in the emerging civil rights movement. The NAACP provided the essential vehicle for meeting the escalation of black expectations and militancy that accompanied the war. NAACP membership in the South by the late 1930s was slightly more than 18,000. By the end of the war it approached 156,000.[36]

That jump in the association's membership owed heavily to the good work of the group, including its legal defense work, its efforts to remove impediments to the black vote, and its southern speaking and recruitment tours featuring prominent national spokesmen like Houston and Marshall. It likewise owed significantly to the indefatigable efforts of Ella Baker, wartime southern field secretary for the NAACP.

While the national office paid much lip service to the notion of making the NAACP relevant to the masses of black people, within the upper echelon an elitist and top-down vision of black liberation dominated. The NAACP leaders firmly believed that they would lead their people to freedom. Baker, however, advanced a far more democratic and participatory vision of black insurgency. She saw herself as a facilitator of local-based movements, working with a broad spectrum in local communities to articulate clearly both common goals and viable strategies for effective collective struggle. In other words, she advanced a bottom-up approach to organization. Baker "spent six months of each year in the South, taking the NAACP to churches, schools, barbershops, bars, and pool halls," writes Sullivan, adding that Baker "helped to build chapters around the needs and concerns of individual communities and encouraged cooperation with labor unions and other progressive organizations."[37]

Baker's emphasis on alliances constituted another article of faith within the NAACP. For example, there was the Southern Negro Youth Congress (1937–48), a group committed to forging links between workers and southern black youth. Similarly, the 1940s South Carolina Progressive Democratic Party constituted another element of the growing black insurgency. Organized labor, notably CIO unions and the Highlander School, with its commitment to working toward interracial labor activism in the South, played crucial roles in fostering support for the black struggle. So did the Communist Party—especially prior to the widespread postwar anti-Communist hysteria and repression. Also important were many New Deal–inspired southerners and white-dominated interracial organizations like the Southern Conference for Human Welfare. In various and sundry ways, these organizations and many other groups and individuals contributed to the groundwork for *Brown*. These were often difficult yet heady times; the 1930s and the pre–cold war 1940s were ultimately, as Sullivan demonstrates, "days of hope."[38] *Brown* was clearly a product of that hope.

THE GROWING ANTI-RACIST OFFENSIVE: *AN AMERICAN DILEMMA* CONFRONTS WORLD WAR II

Another vital development fueling the NAACP's crusade was the declining intellectual and cultural respectability of racism. In *Brown* and the various cases the NAACP lawyers argued leading up to it, the growing scientific and humanistic consensus in favor of egalitarianism was crucial. Nowhere was this point more effectively put forward to national and worldwide audiences than in Gunnar Myrdal's magisterial study of race relations in the United States, *An American Dilemma* (1944). The Swedish economist directed a large staff in an exhaustive study, four years in the making, of the evidence and significance of the discrepancy between the American creed and the American reality for African Americans. The awesome final product consisted of more than 1,000 pages of text, ten appendices, and more than 250 pages of notes (see p. 102 in this book).[39]

For 1950s America and beyond, the *Brown* decision and *An American Dilemma* constitute twin pillars in the evolving liberal racial orthodoxy: America had no choice but to live up to the American creed in its treatment of its black citizens. Evidence of the impact of *An American Dilemma* can be seen in its extensive use in the theory and practice of civil rights law—where its findings became crucial—and its influence on the Supreme Court that decided *Brown*. It became the authoritative work on black-white race relations until the mid-1960s when its assimilationist and integrationist approach came under attack (notably within the black insurgency) as being too liberal, too reformist, and complicitous in the negative construction of black life and culture. From World War II up to the radical Black Power movement beginning in 1966, *An American Dilemma* defined the liberal orthodoxy on American race relations. The *Brown* decision experienced a similar path.

As the antisegregation documents for the period after 1944 in this collection make clear (see chapter 4), the authority of *An American Dilemma* was constantly invoked, implicitly as well as explicitly. The earlier documents demonstrate, moreover, that an understanding of the basic problem discussed in *An American Dilemma*—the disjunction between the

American creed and the white oppression of African Americans—goes back to the nation's founding. Of course it can be traced back even further to the European enslavement of Africans in the New World. Even the related emphasis in Myrdal's text on the baneful impact of white racism on whites as well as on blacks is a recurrent historical theme, traceable here in both earlier and later documents.

As a social scientist committed to moral exhortation and social engineering, Myrdal emphasized both vigorous government leadership and strong government intervention to resolve the problems among blacks engendered by racial prejudice and discrimination. Those engaged in the black freedom struggle—including antebellum abolitionists, postbellum supporters of Reconstruction, and New Deal–inspired racial activists—have shared Myrdal's faith in an activist government committed to racial equality. Unfortunately, at midcentury this activist approach had not found enough public support.[40]

While there was much new and original material in *An American Dilemma,* what was particularly striking then and now is how well the text captured the evolving liberal support of racial egalitarianism and integrationism among the lay public and scholars, especially sociologists and anthropologists. Myrdal employed many of the best available black and white minds for his study and distilled the results of their contributions through his own perspective as a relative outsider to the American scene. The fact that extraordinary national effort had to be undertaken to ameliorate the inequalities African Americans experienced was patently clear. In line with its scholarly and objective goals, though, Myrdal's text was very long on description and analysis and short on policy prescriptions.

As in *Brown,* the argument and the remedy in *An American Dilemma*—like most American efforts to deal with racial inequality—did not go far enough. What became increasingly clear in the period from *An American Dilemma* to *Brown* was a growing yet insufficient national will to tackle this thorny problem. In spite of brief moments to the contrary, such as the noteworthy government efforts spawned by black insurgency between 1954 and 1974, the national will has proven insufficient to the challenge.

Even the explosive wartime economy that brought the nation out of the Depression and the subsequent thirty years of sustained economic growth were insufficient to create racial equality. Neither was postwar U.S. global supremacy. Nonetheless, in this broad context of sustained economic growth and "Pax Americana," or worldwide U.S. dominance, the black freedom struggle surged. *Brown* represented a turning point in its building momentum.

The pulsating wartime economy transformed the American landscape. Streams of rural blacks leaving the South during the Depression reached flood proportions during the war as job opportunities and prospects for a better life proliferated in northern and western cities. Heightened black political consciousness engendered by the Depression continued to grow during the war. Increasingly, the race problem became a national issue, not merely a southern one. As Pax Americana demanded that the United States assume the awesome pressures and glaring spotlight of international center stage, African Americans and their allies fully understood that from a geopolitical perspective, state-enforced white supremacy was indefensible. In this radically altered context, with its local southern black membership base expanding and energized, the NAACP shifted its strategic attack from equalization to direct attack.

This significant shift reflected several developmental and organizational factors as well. In the South, there were increasing numbers of blacks willing to file civil rights cases and black lawyers able to argue those cases. The NAACP legal staff had grown in size and maturity, reaching the point where by the war's end it had become a well-oiled and flexible machine. In 1939 the NAACP created the NAACP Legal Defense and Educational Fund as a functionally autonomous wing. This streamlined and enhanced the association's legal enterprise. By 1945 the staff had coalesced around the move from equalization to direct attack and in 1948 the board of directors and the Annual Conference issued a full-fledged statement in support of the direct attack strategy.[41]

Recent court successes by the NAACP legal team and its cohorts, especially the 1944 Supreme Court ruling in *Smith v. Allwright* outlawing the white primary in the South spurred that support.[42] This racially exclusionary device had functioned as a critical prop of white, one-party, Democratic rule in the South. In a related vein, a significant measure of the success of NAACP organizing in the 1940s owed to increasing black political mobilization around voting, particularly in the South. Growing black political power in the North enhanced the national impact of black politicization in general. Chicago's South Side and New York City's Harlem, where recently elected black congressmen were beginning to flex their political muscles—most notably Harlem's Reverend Adam

Clayton Powell Jr. — signaled this important trend. In the overall domestic and international context, an all-out legal assault against Jim Crow and *Plessy* became a viable enterprise.

CONTINUITY AND CHANGE IN THE LEGAL STRUGGLE: EQUALITY, EQUALIZATION, AND DIRECT ATTACK

The creation of legal precedents was absolutely essential to the NAACP's overall strategy against Jim Crow. Reverses as well as victories thus proved to be invaluable learning tools. Indeed the long-term "road to *Brown*" had many ups and downs.[43] This collection includes two key legal setbacks in the nineteenth century: *Roberts v. City of Boston* (1849), which dealt with equal educational opportunity (see p. 42); and *Plessy v. Ferguson* (1896), which legitimized separate but equal railway accommodations. The former case painfully revealed that the prejudice and discrimination endured by free blacks in the antebellum slave South had clear parallels in the free states of the antebellum North. White supremacy was a national dilemma, not a regional one.

Roberts was a Supreme Judicial Court of Massachusetts decision that separate common or public schools for Boston's black schoolchildren did not deny them their legal rights and did not expose them to undue logistical difficulties or degradation. In addition, Massachusetts' highest court agreed with the defendant, the Boston School Committee, that it was within its constitutionally delegated power to separate black schoolchildren from white schoolchildren, given the committee's statutory authority over the ways and means of local public education. If in the committee's judgment racially segregated schools served reasonable educational and sociopolitical objectives, the ruling maintained, then this particular form of racial discrimination was legal. While a state law in 1855 overturned the original decision, the resonances between the *Roberts* and *Plessy* cases and the significance of both cases in the history of American apartheid are revealing.[44]

An even more stunning and influential constitutional setback was *Plessy v. Ferguson*. The majority opinion cited *Roberts v. City of Boston* as one among several key precedents. The legal logic in *Plessy*, as in *Roberts*, owed heavily to social customs rooted in white supremacy. It also relied on an interpretation of the Fourteenth Amendment's guarantee of each citizen's right to equal protection under the law as consistent with racially separate but equal public accommodations and institutions. In later cases,

Plessy was at times interpreted narrowly as affirming segregation in transportation and comparable kinds of public accommodations, while *Roberts* affirmed segregated education. As the documents demonstrate, pro-segregation legal cases relied extensively on these distinctions and related arguments and rulings.

In the case for the black plaintiff in *Roberts,* the venerable Massachusetts abolitionist and senator Charles Sumner eloquently articulated an elaborate and powerful brief for the concept of racial equality as well as the policy of integrated public schools in "enlightened" Boston. Again, the pro-egalitarian and pro-integration documents here demonstrate that the twentieth-century "road to *Brown"* made extensive use of Sumner's stirring and ultimately compelling nineteenth-century brief. In many ways, the antebellum abolitionist crusade that gave rise to Sumner's brief later reconfigured itself into a neo-abolitionist crusade against Jim Crow. The NAACP's legal campaign exemplified this transition.[45]

Justice John Marshall Harlan's famous dissent in *Plessy* is best known for its articulation of the Constitution as color-blind. He wrote that "in view of the Constitution, in the eye of the law, there is in this country no superior, dominant, ruling class of citizens. There is no caste here. Our Constitution is color-blind, and neither knows nor tolerates classes among citizens. In respect of civil rights, all citizens are equal before the law." This inspiring and idealistic vision was eventually enshrined in *Brown,* furnishing the egalitarian and integrationist forces with a powerful endorsement.

Less well known and less often discussed was Harlan's embrace of de facto white supremacy and his opposition to any kind of social equality between the races. As seen in documents here, Jim Crow's legal partisans often quoted Harlan on these points as a way of undercutting his assertion that the Constitution is color-blind. Harlan's acceptance of segregated public school education as consistent with state power likewise curried favor with advocates of Jim Crow education and incurred the opprobrium of its opponents. His dissent proved to be very influential in large measure precisely because of its double edge.[46]

Until direct attack became the NAACP's guiding strategy in the late 1940s, both sides accepted the *Plessy*-defined terms of the debate— separate and equal—as the controlling issue. *Gong Lum v. Rice* (1927) illustrates how *Plessy* carried the day. The father of nine-year-old Martha Lum sought admission for his daughter to a local white school in Mississippi on the grounds that his family was of Chinese descent. He argued that it was wrong for Martha to be compelled to attend the black school, given the stigma attached to blacks and their separate schools, especially as his daughter was not black. Neither was she white, the Supreme Court argued, as it upheld the power of the state to categorize and place students as it saw fit. The issue here was not the right of the state to maintain segregated schools, which the plaintiff accepted. Rather, the issue was both legal and categorical: the state of Mississippi could compel the girl to go to a black school when she was neither black nor white, but of Chinese descent.

This case is also instructive in its erasure of Chinese racial identity and its conflation of that identity with a black racial identity. Two points, among others, are critical to this discussion. First, the dualistic construction of race in America obscures both powerful cultural differences among nonwhites, in this case blacks and Chinese, and critical differences in their historical experiences in the United States. This racial dualism also misrepresents and thus devalues the integrity of their group-based identities. In turn, it buttresses white supremacy.[47]

Second, it is worth thinking critically about the power of the state, or the government, to determine racial identities or to define who belongs to which race. That power did not reside ultimately with the oppressed, nonwhite minorities themselves, in this instance with the black and Chinese citizens. Indeed, a vital aspect of the Asian American movement, particularly between the late 1960s and early 1980s, and the black civil rights and Black Power movements (1955–75) was the same. Both fought to wrest the ultimate power of group definition from the state and federal governments and to reassert control over their identity: to define on their terms who they are.[48]

In the years before *Brown,* the search for precedential decisions undermining *Plessy* proceeded. In the area of equal educational opportunity for blacks, particularly in the Upper South, the NAACP carved out a series of important victories in salary equalization cases in the 1930s. The larger strategic problem, of course, had been anticipated: the process was piecemeal, gradual, and very time-consuming because each separate school jurisdiction had to be challenged separately. In addition, the tactical move among southern school districts to mask racial disparities in salaries through the introduction of merit criteria exposed a serious flaw in the salary equalization strategy and pushed the NAACP lawyers toward the direct attack strategy.[49]

Similarly, a series of "victories" in graduate and professional school cases—*Pearson v. Murray* (1936), *Missouri ex rel. Gaines v. Canada* (1938), *Sipuel v. Oklahoma State Regents* (1948) — nibbled away at *Plessy.*

Taken together and over time, these cases played out the equalization approach to the point where the direct attack approach became imperative. In *Pearson v. Murray,* the Maryland Court of Appeals ruled that Donald Murray, a black Amherst College graduate, had been denied equality of educational opportunity when he was refused admission to the University of Maryland's law school. The state's alternative of providing scholarships for blacks to attend out-of-state schools was viewed as a violation of Murray's Fourteenth Amendment right to equal treatment under the law. Because the constitutional injury to Murray was "present and personal," the remedy had to be immediate. Murray either had to be admitted at once to Maryland's School of Law or a separate and equal school of law for Maryland blacks had to be created forthwith. Since a comparable black law school could not be created overnight, he had to be admitted to Maryland's School of Law.

Nevertheless, with the possibility that a separate black law school might satisfy the letter of the ruling, *Plessy* clearly remained intact. In the *Gaines* and *Sipuel* cases, similar circumstances resulted in similar rulings, this time in the Supreme Court. The decisions in these cases turned on the issue of the inequality between the reputable all-white state-supported law schools in Missouri and Oklahoma and the makeshift all-black arrangements those states scrambled to provide to avoid admitting blacks to their all-white law schools. Notwithstanding the impact of the sociological arguments on the behind-the-scenes discussions of these cases, the Court was deeply divided on the issue of overruling *Plessy* and thus did not go that far.[50]

For the NAACP, the *Gaines* decision was a moment to be savored, however; it was the first favorable Supreme Court judgment casting doubt on the legality of *Plessy.* In a comparable decision in *Sipuel,* the Supreme Court upheld the right of the black plaintiffs to equal educational opportunity, although possibly under separate circumstances. The limits of equalization as a strategy were becoming patently clear, particularly in light of fallacious defense arguments claiming the comparability as against the actual equality of separate schools. By the late 1940s, the NAACP legal trust was working on a Texas case, *Sweatt v. Painter* (1949), where the direct attack strategy was being readied (see p. 87).

The strategy employed in *Sweatt v. Painter* and a related case, *McLaurin v. Oklahoma State Regents* (1949), failed to get the Supreme Court to overturn *Plessy.* Even so, that same strategy would soon prove effective in *Brown.* In the *Sweatt* and *McLaurin* cases, the NAACP lawyers used a two-pronged approach. First, they focused on how separate black law schools, especially quickly contrived ones created to forestall integration, lacked the many advantages of the traditional all-white law schools and were thus a blatant denial of equal educational opportunity. Second, in a tactical innovation, the NAACP lawyers utilized psychosocial evidence on the harm segregation inflicted on its victims. This kind of argument had been used in a 1945 friend-of-the-court brief filed in support of a lawsuit against Orange County, California, for its practice of segregating Mexican American schoolchildren from white schoolchildren. In *Sweatt* and *McLaurin,* this tactic foreshadowed the increasingly influential emphasis in postwar America on the psychological damage that segregation inflicted on blacks.

In *Henderson v. United States* (1949), a case coupled with *Sweatt* and *McLaurin,* the federal government issued a friend-of-the-court brief vigorously condemning segregated railroad dining cars, which the Court subsequently declared illegal. Earlier, in *Shelley v. Kraemer* (1947) and *Sipes v. McGhee* (1947), the Supreme Court outlawed restrictive covenants (contracts forbidding the sale of property to blacks and other "stigmatized" groups and individuals) as invidious and unconstitutional forms of racial discrimination. Charles Houston himself, in concert with his NAACP colleagues, argued this series of cases. In its friend-of-the-court brief to support the government's opposition to restrictive covenants, the Department of Justice revealed the growing importance of cold war concerns. That brief made it clear that Jim Crow was a very serious problem for the United States in its propaganda war with the Soviets for the hearts and minds of the Third World, especially in Africa.[51] Indeed, this was an issue that the NAACP legal team and its cohorts increasingly exploited to good effect.

In *Henderson,* Attorney General Howard McGrath maintained before the Supreme Court that "segregation signifies and is intended to signify that a member of the colored race is not equal to the white race." Jim Crow, McGrath further explained, represented "an anachronism which a half-century of history and experience has shown to be a departure from the basic constitutional principle that all Americans, regardless of their race or color or religion or national origin, stand equal and alike in the sight of the law."[52] This ringing endorsement of constitutional egalitarianism by the nation's number one lawyer meshed well with the Justice Department's earlier argument for desegregation on cold war grounds. This kind of ammunition, including President Harry Truman's official initiation of desegregation of the armed forces, verified strong opposition within the government to state-sanctioned racial segregation. The stage was now set for a full-fledged direct attack against *Plessy*-sanctioned segregation: the "road to *Brown*" was taking shape.

POLITICS, SOCIAL CHANGE, AND DECISION-MAKING WITHIN THE SUPREME COURT: THE CRAFTING OF *BROWN*

Brown v. Board of Education of Topeka, Kansas, as well as *Briggs v. Elliott, Davis v. County School Board of Prince Edward County, Belton v. Gebhart,* and *Bolling v. Sharpe* — the cases eventually argued collectively as *Brown v. Board of Education* — all wound their separate ways toward the Supreme Court in the early 1950s. In each case, and in spite of anticipated lower court setbacks, the NAACP legal staff remained hopeful about a positive Supreme Court ruling in favor of equal educational opportunity. Third World nationalist struggles, most importantly growing assertiveness within America's own communities of color, pervaded the international community, which was reeling from the Holocaust, the dropping of the atomic bomb on Hiroshima and Nagasaki, and an escalating cold war. Worldwide as well as at home, white supremacy was under furious assault. Even though the "Red scare" repressed left-progressive forces in this country, seriously undermining the most radical elements within the black struggle, that insurgency soon reinvigorated itself via the civil rights movement. *Brown* contributed significantly to the ethos and spirit of this revitalized social movement, which was fast becoming a mass movement.

As the excerpts collected here show, both sides in *Brown* mounted strong cases. From the lower courts, *Briggs v. Elliott* (see p. 126) is included because the case featured two legal titans: the celebrated establishment lawyer John W. Davis for the defense and Thurgood Marshall for the plaintiffs. In oral arguments, they both provided high drama as well as astute argumentation. In their legal briefs, they compellingly presented their cases. *Briggs v. Elliott* encapsulated the twin battles in the NAACP's all-out war on segregated schools. First was the clear-cut evidence of the denial of equal educational opportunity owing to gross physical and funding disparities between white and black schools. Second was the interrelated argument of psychosocial harm inflicted on black schoolchildren as a result of Jim Crow schools. Although only the lower court dissent of Judge J. Waties J. Waring responded favorably to the second argument, it clearly made an impact on both sides.

Indeed, the sociological argument figured in all of the component cases in *Brown* except *Bolling v. Sharpe*. The NAACP lawyers relied heavily on the social-scientific work of many influential scholars such as Otto Klineberg and Gordon Allport. The most important of these experts, however, were social psychologists Kenneth and Mamie Clark. The Clarks had devised a doll test as a way to gauge evidence of personality dysfunction among black children under Jim Crow. When shown two dolls—one white and one black—the children were asked which one they preferred. The fact that a preponderance of the black children expressed a preference for the white doll was most revealing for the Clarks. From this finding, they extrapolated that the damage done to the self-esteem of these children reinforced notions of black inferiority and white superiority. Racial segregation did indeed damage the black psyche. The issue was not, as the majority opinion in *Plessy* had contended, that the antiblack stigma was all in the minds of blacks. Rather, the stigma was all too real, for whites as well as blacks.[53]

In the lower court opposition, only in *Davis v. County School Board of Prince Edward County* did the defendants use experts to rebut the sociological argument. Throughout the lower court litigation in these cases, however, the sociological argument forced the opposition at least to seek to diffuse it. In the Topeka case, the defendants questioned the social-scientific viability of the sociological evidence and its specific applicability to the children under consideration as opposed to black schoolchildren in Jim Crow schools generally. In other words, except for the Clarendon County data offered by the Clarks, the evidence in these cases was often drawn from studies conducted outside the community in question.

At the time and subsequently, this sociological jurisprudence spawned innumerable critics as well as supporters. Many in both camps bemoaned the sociological incursion into legal argumentation. Supporters have stressed that the evidence of the deleterious impact of racial caste on the American psyche, in particular the black psyche, could not be diminished by reference to the methodological and interpretive limitations of the social psychological evidence—especially the Clarks' doll test. Instead, they have argued, on balance the Clarks' study was persuasive. Furthermore, for many, the gravity of the constitutional and moral issues involved overrode considerations of both scholarly detachment and judicial restraint, compelling intellectual and legal activism.

The right-wing conservative opposition blasted what they saw as blatant social engineering. They strongly decried liberal and integrationist bias masquerading as legal reasoning and reiterated their opposition to sociological jurisprudence and their support for judicial restraint and scholarly detachment. Other opponents, among them racial moderates and liberals, criticized the social-scientific evidence as more faddish than substantive, more ambiguous and contested than clear-cut and persuasive. Some pointed out problems in research design, methodology, and assessment of evidence.

Some critics of the Clarks' doll test highlighted the need to disentangle the influence of the broader environment from that of the school. It was also argued that black children might identify with nonblack, even white, images without being imprisoned by self-hate. Some argued that especially outside the South, similar kinds of tests have shown that close proximity to whites, or integration, yielded comparable evidence of black self-hate. For some, it was a question of what was worse for black self-esteem, integration or segregation? In fact, social science experts from 1919 to 1941 downplayed the argument of lowered black self-esteem as a principal consequence of segregation, notably of all-black institutions like schools. These experts saw the problem of black self-esteem as a more complex phenomenon. Nevertheless, between the *Brown* decision and the late 1960s black challenge to integrationism, an expanding social-scientific consensus about the negative effects of segregation on the black psyche became increasingly important to the argument for integration.[54]

Chief Justice Earl Warren's reference to the persuasiveness of this sociological evidence in the very text of the 1954 decision touched off a vigorous debate. Footnote 11 of the decision lists seven authorities — an address by Kenneth Clark leads off — and draws directly from the NAACP brief. This famous footnote further heightened the controversy around that evidence and the decision itself.[55]

Many, particularly judicial conservatives and pro-segregationists, saw this use of sociological evidence as deeply threatening. They viewed the decision as a judicial usurpation of states' rights and of federal legislative and executive powers. It represented a flagrant abuse of judicial review: the right of the Supreme Court to rule on the constitutionality of the acts of the other branches of government.

They also saw the decision as part of an all-out assault on white supremacy, or, euphemistically speaking, southern mores. This kind of thinking led to an extremist southern white backlash to *Brown* and integrationism: a movement identified as "massive resistance." Included here is the "Southern Manifesto," the foundational document of this racist and reactionary white counterinsurgency (p. 220). Signed by ninety-six southern white leaders, this document pledged unyielding opposition to judicial usurpation and dedication to overturning the decision.

This kind of response had been feared by the justices, several of whom were southerners, because they clearly perceived the depth of white racist attachment to Jim Crow. In fact, by the time *Brown* reached the high court in December 1952, a majority of the justices were predisposed to striking down *Plessy*. Under the failed leadership of Chief Justice Fred Vinson, however, the Supreme Court had been deeply fractured and unable to forge a consensus about overturning *Plessy*. Changes in court personnel, the most important of which was the death of Vinson in September 1953 and the selection of Earl Warren as chief justice, proved critical. Warren and his colleagues understood that the enormity of overruling a precedent of *Plessy*'s magnitude demanded that the decision be unanimous. A divided ruling would dilute its impact.

The arguments in *Brown* proceeded in three stages. First, the initial presentation of the case did not give the justices all the time and evidence they needed to decide definitively. To gain more time to sift through the evidence and try to sway one another on various points, the justices called for a second stage to the proceedings: reargument on the intentions of the Fourteenth Amendment's framers. Had the framers created that amendment as opposing or supporting segregated public school education? The third and final stage, after the decision to strike down *Plessy* on the merits of the case was reached—often called *Brown I*—the court called for arguments about the remedy, or how to enforce the ruling. That decision regarding implementation is often referred to as *Brown II*.

During the first stage of deliberations, a consensus emerged that overall the claimants' case was powerful enough to be sustained and in turn to be used as a platform to overrule *Plessy*. Speaking to the sociological argument, Justice Tom Clark privately observed that "we need no modern psychologist to tell us that 'enforced separation of the two races stamp[s] the colored race with a badge of inferiority,' contrary to [the argument in] *Plessy v. Ferguson.*" Agreement on racial equality and the related imperative of equality of educational opportunity emerged early on. Two other issues loomed as more contentious. In spite of much debate, the evidence of the intentions of the Fourteenth Amendment's framers was not very convincing. A modest preponderance of the evidence suggested that they saw the amendment as favoring segregated public school education. Ultimately, however, the justices found the evidence inconclusive and wholly insufficient to sustain a judgment one way or the other.[56]

There was also much interesting behind-the-scenes debate about whether the case could be decided principally on its legal merits or whether political and social considerations were primary. Once the focus shifted away from the traditional issues of original intent and reliance on precedent to considerations of the impact or consequences of an admitted error—*Plessy*—the die was cast. As Mark Tushnet demonstrates in the reasoning of Justice Robert Jackson, "an appropriate premise for overruling *Plessy*" did not necessitate compelling explanations of either "the failure of the representative branches [Congress and the president] or the intentions of the framers." Instead, the appropriate legal premise was a profound mid-twentieth-century global paradigm shift: an emerging and increasingly powerful consensus regarding racial equality. Within a legal framework based on this premise rather than the fallacy of white racial superiority, the view of racial equality as a fundamental principle of law as well as of society and culture meant that state-sanctioned racial segregation in education, and beyond, was a dead constitutional letter.[57]

A consensus within the Court about overruling *Plessy* was easier to reach than an agreement regarding remedy. Indeed, much of the debate among the justices about overturning *Plessy* pivoted around how, in effect, to implement such a potentially cataclysmic decision. This deeply felt sensitivity about how the nation, especially the white South, would react clearly circumscribed the whole of the Court's lengthy deliberations, not just the third stage where remedy was the explicit subject. The justices, like many Americans, feared what Justice Clark prophetically referred to as "subversion or even defiance of our mandates in many communities." In an early closed conference of the justices on the case, Alabama-born Justice Hugo Black had pointed to the issue agitating all of the justices: the extraordinary depth of racial caste in the South, the "deep seated antagonism to commingling" across racial boundaries. He argued that many southern school districts would shut down "rather than mix races at grade and high school levels." Ultimately, however, the issue of the differences between desegregating at the level of colleges and universities as opposed to the primary and secondary levels, where social intercourse between the races was seen as far more explosive, did not prove determinative.[58]

What did prove compelling was agreement that the remedy had to be gradual. Without this commitment to incrementalism, the commitment to overturning *Plessy* weakened. The contending briefs on this issue fully aired both sides: immediatism and gradualism (see p. 175). Fears of an extremely volatile southern white response to an immediate implementation decree rendered gradualism the only viable option. At the point of remedy, therefore, legal concerns were plainly secondary. The NAACP brief argued strongly for immediate relief given that egregious violations of constitutional rights had been established. But the Court wanted to weigh the effects of immediate or gradual implementation. This delicate situation led to the ambiguous and in many ways ill-fated compromise of the eventual relief decree: implementation "with all deliberate speed."[59]

The Court's fundamental lack of nerve and will mirrored that of the executive and congressional leadership as well as of the vast majority of white Americans. As Tushnet has shown, "It was not so much that the Justices understood that it would be difficult for courts to accomplish what they wanted through judicial decrees: the more acute problem was that they never truly decided what they wanted the courts to accomplish."[60]

While they hoped and prayed for the best, the NAACP lawyers and perceptive observers everywhere were fully aware that the relief decree lacked muscle. A less radical and perhaps more effective decree might have been the middle ground option—immediate desegregation tempered by modifications sensitive to local conditions. On the cusp of the twenty-first century, the continuing national scandal of separate and unequal schools for children of color is a tragedy of epic proportions. The same is true of the persistence of racial apartheid in many areas of American life, including housing and employment. Tushnet has provocatively offered in retrospect that "had the Court followed through on the promises of *Brown,* political resistance to desegregation might have been smaller, the courts might not have had to develop intrusive remedies, and the reaction against 'judicial activism' . . . might not have occurred." Perhaps the "shock therapy" of immediatism would have served the short-term future better.[61] It is hard to imagine it serving worse than "all deliberate speed."

THE *BROWN* DECISION: IMMEDIATE RESPONSES AND IMMEDIATE CONSEQUENCES

As soon as Chief Justice Warren announced the decision on May 17, 1954, the reaction was swift and predictable. Most commentators did not emphasize the exceedingly moderate and measured language of the decision. The legal rhetoric neither soared nor inspired. The text stressed that segregation is wrong and had damaged all Americans, especially blacks. It also emphasized that in midcentury America, Jim Crow was morally and intellectually indefensible. There was no ringing rejection of segregation as a profound legal error in *Plessy.* There was no pointed legal argument against race as an arbitrary and thus indefensible category or for a color-blind society. The egalitarianism of the decision's text was dutiful and restrained.[62]

Far more important than the modest substance of the text of the decision has been its awesome symbolic resonance—what Americans have read into the decision, how they have interpreted it. Blacks and their allies in the black liberation struggle were pleased and hopeful, but often cautious. Pro-segregationists and their allies were deeply alarmed. Both camps fretted about the immediate and long-term consequences of this momentous decision. In particular, racial liberals and supporters of the black freedom struggle felt vindicated. Richard Kluger writes that the decision "represented nothing short of a reconsecration of American ideals. . . . The Court had restored to the American people a measure of humanity that had been drained away in their climb to worldwide ascent." Not surprisingly, therefore, within an hour of the decision's announcement, "the Voice of America would begin beaming word to the world in thirty-four languages: In the United States, schoolchildren could no longer be segregated by race."[63]

Nationwide editorial comment reflected a predictable range of opinion (see p. 199). Notwithstanding an undercurrent of caution, black newspapers throughout the country lauded the decision as heralding a new age in race relations. The white press more clearly reflected local and regional perspectives. Northern and western newspapers saw the ruling as positive and hopeful. Pro-segregationist southern papers uniformly condemned the decision, while the liberal southern white press summoned up a guarded hope for the best.

Black response ran the gamut from elation to occasional opposition. Cleophus Brown, a labor and civil rights leader in Richmond, California, remembered that moment as "the point at which 'black folks in Richmond saw the light' and really began to believe they could break through." Black cultural racialists like anthropologist and writer Zora Neale Hurston rejected the logic of *Brown* as self-defeating at best, antiblack at worst. Reflecting a tactical position uncommon among blacks, she chose to emphasize the strengths of all-black institutions rather than the inequities under which they labored. From Hurston's perspective, the decision plainly reiterated notions of black inferiority, with its insinuation that black schoolchildren could learn best under the tutelage of white teachers, sitting next to white students. For Hurston, this was a brutal slap in the face of black teachers and administrators as well as black schoolchildren. She charged: "How much satisfaction can I get from a court order for somebody to associate with me who does not wish to be near me?"[64] In fact, a critical failure of the egalitarianism of the liberal and social-scientific consensus undergirding *Brown* was its devaluation of black culture and black institutions and, ultimately, of blacks themselves.

While W. E. B. Du Bois lauded *Brown,* he shared Hurston's concern about its limitations and possible consequences. He was especially troubled by the decision's blindness both to the potential for the mistreatment of black children in integrated schools and to the strengths of a distinctive black culture. Still, on balance, he viewed *Brown I* and *II* as important but imperfect steps along freedom's bumpy journey.[65]

Nevertheless, the legacy of *Brown* rightly looms large in interpretations of the modern American experience. *Brown* is peerless as a moral touchstone in the legal struggle for African American rights. H. M. Levin contends that *Brown* "was central to eliciting the moral outrage that both blacks and whites were to feel and express about segregation, and this new awareness set the stage for the changes that were to follow." In 1957 Albert Blaustein and Clarence C. Ferguson characterized *Brown* as "the most controversial and far-reaching decision of the Twentieth Century." *Brown* "has had a greater impact upon American life than any other legal decision in our history," they wrote, "and it will remain a source of contention and commentary for generations to come."[66]

Brown's jurisprudential legacy is equally impressive, albeit a bit more contested. As Tushnet observed in 1991, "for nearly forty years, *Brown v. Board of Education* has defined the central values of constitutional adjudication in the United States." More specifically, it has contributed enormously to subsequent civil rights battles and social movements on behalf of the marginalized, including other peoples of color, women, gays and lesbians, and the disabled. Likewise, the decision has profoundly influenced the evolution of "rights consciousness" within American society—that is, "judicial activism on behalf of human rights," notably the rights of oppressed groups and individuals. It constitutes a precedential bulwark in the burgeoning legal fields of human rights law, public interest law, and civil rights law.[67]

The pro-egalitarian and antiracist minority on both the right and the left who question *Brown*'s impact largely pursue two lines of argument. Some, like Du Bois, question the effectiveness of the law and the courts as a central arena for the pursuit of social reform. They interpret the post-*Brown* realities of continuing segregation in schools (and beyond) and the related persistence of white supremacy in many forms as proof of the weaknesses of *Brown*. Conservatives use this evidence to underscore their opposition to judicial activism and social change, especially legally mandated. Radicals use the same evidence to call for more effective kinds of judicial activism, as well as social change via law as well as other avenues.[68]

It should be clear by now, however, that neither *Brown* nor the judiciary can be summoned to resolve America's complex and persistent race problems. As this discussion has emphasized, amelioration of this national dilemma has demanded yet failed to generate national will, commitment, and action. The courts are particularly impotent in this regard as they constitute the nonrepresentative branch of government. They have no real power to institutionalize racial equality, not to mention racial integration. Seeking to achieve racial equality principally through legal as opposed to economic, political, and social means has contributed to what the legal scholar Morton J. Horwitz terms a "distortion" in "the battle against racial discrimination." He explains: "The schools—the weakest and most vulnerable of American institutions—have been forced to bear the brunt of the social change required in the battle against racial discrimination, even though school segregation is now largely a function of discriminatory housing patterns which are, in turn, related to job discrimination."[69]

The distortion engendered by the "legalization of the problem of racial discrimination" is largely a post-*Brown* phenomenon. Horwitz astutely observes that "our legal system is overwhelmingly geared to a conception of redressing individual grievances, not of vindicating group rights or of correcting generalized patterns of injustice. This perspective does not easily encourage judges to focus on the burdens, stigmas, and scars produced by history."[70] This bias favoring the individual has undercut the courts' limited ability to alleviate group-based patterns of racial inequality.

In a related vein, the courts' conservative, at times reactionary, refusal to grapple with the clear-cut links between economic and racial inequality exacerbates the disparity between *Brown*'s promise of equal opportunity and the post-*Brown* reality of unequal opportunity for blacks. In the end, the salient issue remains that courts view socioeconomic inequality and racial inequality as separable and in crucial ways unrelated.[71]

In 1979 Horwitz outlined an extensive and telling series of basic dilemmas *Brown* has presented over time. At that point, twenty-five years after the decision, these dilemmas remained unresolved.

Does it stand simply for color blindness—for the principle that it is constitutionally impermissible for the state to take race into account even for benign purposes—or instead does it stand as a barrier only to the use of racial classifications for the purpose of oppressing minorities? Should the principle of *Brown* continue to be directed only at governmental discrimination—so-called state action—or should it apply to private action as well? Does *Brown* require only that racial minorities be provided equality of opportunity? But what happens when even after all of the formal barriers of exclusion are dropped, the intangible culture of racism or the scars of a history of deprivation continue to produce racially unequal consequences? Is a racially discriminatory program one that is intended to produce unequal results or one that actually produces such results regardless of the intentions or motivations of its creators? Do such programs interfere with the constitutional rights of non-minority members who may be excluded because of minority preference for jobs, housing, or admission?[72]

At the end of the twentieth century, almost fifty years after *Brown,* these questions are still pressing. Horwitz's structural critique of liberal jurisprudence recalls Du Bois's critique of the NAACP's legalism in the

1930s. At that time Du Bois had come to embrace black economic nationalism within the confines of voluntary segregation. The key point of both arguments is compelling: legal change can obviously accomplish only so much without fundamental economic change.

Similarly, the radical 1930s politics of Du Bois and the conservative 1950s politics of Hurston converged around a shared critique of the cultural politics shaping the road to and beyond *Brown*. Both Hurston and Du Bois saw the negative views of black culture in the liberal assimilationism driving the NAACP legal assault against Jim Crow as wrongheaded and dangerous. Unsuccessfully, they furiously argued against it. The NAACP legal campaign had pushed black integration into the white American mainstream, a norm against which African American culture was measured and found woefully wanting.

The assimilationist view of African American culture as defective was deeply political. In fact, it was far more political than anthropological. Those like Thurgood Marshall and highly influential black sociologist E. Franklin Frazier who argued that black culture was in crucial ways flawed and dysfunctional did so in large part because they wanted to place all of the onus for the African American situation on white oppression of blacks. From their point of view, the issue was black powerlessness: Africans had been stripped of their African cultures. They had lost their Africanness. They were Americans, yet racially oppressed and therefore marginalized.

This blatantly political view of black culture thus had a pointed goal. Its proponents, black and white, wanted none of the blame for the "black condition" to be loaded off onto the "Africanness," or blackness, of black people. White Americans had to be made to feel responsible for their racist oppression of blacks and galvanized in the process to alleviate it. If the responsibility for the "black condition" could be foisted onto the racial and cultural distinctiveness of blacks, then whites might be able to blame black problems on those differences, not on white racism.[73]

Unfortunately, as Du Bois and Hurston argued, in its erasure of both the enduring Africanness and related strengths of black cultural uniqueness, this brand of liberal racial politics demeans blackness. Indeed, in a related and revealing context, historian Daryl Michael Scott has demonstrated that the modern political manipulation of "the image of the damaged black psyche" has been both invidious and widespread. He shows that what he characterizes as

> damage imagery has been the product of liberals and conservatives, of racists and antiracists. Often playing on white contempt towards blacks, racial conservatives have sought to use findings of black pathology to justify exclusionary policies and to explain the dire conditions under

which black people live. Often seeking to manipulate white pity, racial liberals have used damage imagery primarily to justify policies of inclusion and rehabilitation. Even when relatively devoid of emotional appeals or damage, the social science image of black personality has historically been sketched by experts motivated or heavily influenced by racial ideologies and politics.[74]

A cultural analogue of the "damaged black psyche" has been the concept of black cultural inferiority. The false and misleading linkage of psychological damage with cultural deficiency has reinforced the notion of black inferiority as both innate, or biological, and improvable, or environmental. Even though biological concepts of race are scientifically indefensible and cultural concepts of race are often equally erroneous and misleading, both persist. Indeed, the tenacious myth of black inferiority typically blurs the distinction between biology and culture. In the popular American imagination, cultural and racial inferiority are seen as synonymous, playing into the notion of black inferiority as natural. Regrettably, these (mis)understandings obscure the fundamental fact of egalitarianism—we are all far more alike than different—and continue to haunt post-*Brown* America.

Nevertheless, *Brown* deserves to be recognized for its enormously liberating impact on America and the world. Post-*Brown* American society was forced to look deep within itself and confront the fundamental problem of white racism and its impact on whites and blacks alike. As the great African American leader Frederick Douglass observed in the nineteenth century, the American race problem is essentially a white problem. Structural inequalities bracing white privilege in concert with white racist notions of black social pathology, black cultural inadequacy, and black biological inferiority have fueled America's historic refusal to grapple with the real problem: white racism. *Brown* was a wake-up call America continues to struggle with.

Reflecting unconditional faith in the best of the founding American ideals, *Brown* signifies hope for America's future. It stands for a better America: a humane, inclusive, and free America. The profundity of that vision propelled the black freedom struggle of which *Brown* was a vital part. In 1965 Martin Luther King Jr. paid homage to the centrality of the legal campaign against American apartheid. "The road to freedom," King observed, "is now a highway because lawyers throughout the land, yesterday and today, have helped clear the obstructions, have helped eliminate roadblocks, by their selfless, courageous espousal of difficult and unpopular causes."[75]

White resistance to black equality and empowerment has historically been fierce, and the reaction to Brown was no different. In August 1955, three months after *Brown II* had been announced, Emmet Till, a fourteen-year-old black teenager visiting relatives in Mississippi, was lynched for allegedly whistling at a white woman. In December of that year blacks in Montgomery, Alabama, launched the successful year-long Montgomery bus boycott. In September 1957 President Dwight D. Eisenhower was forced to send federal troops to Little Rock, Arkansas, to protect black schoolchildren integrating previously all-white Central High. The conclusion of Cyrus Cassell's poem "Soul Make a Path Through Shouting" poignantly captures the riveting drama and complex historical context of that most revealing moment:

> I have never seen the likes of you,
> Pioneer in dark glasses:
> You won't show the mob your eyes,
> But I know your gaze,
> Steady-on-the-North-Star, burning—
>
> With their jerry-rigged faith,
> Their spear of the American flag,
> How could they dare to believe
> You're someone scared?;
> *Nigger, burr-headed girl,*
> *Where are you going?*
>
> *I'm just going to school.*[76]

Brown gave us that heroic moment and infinite others. Most important, the struggle to realize the promise of *Brown* endures.

NOTES

[1]Richard Kluger, *Simple Justice: The History of Brown v. Board of Education and Black America's Struggle for Equality* (New York: Knopf, 1975), 4, 6, 8, 14.

[2]Ibid., 3–26.

[3]Paul Finkelman, *Dred Scott v. Sandford: A Brief History with Documents* (Boston: Bedford Books, 1997).

[4]Robert C. Toll, *Blacking Up: The Minstrel Show in Nineteenth-Century America* (New York: Oxford University Press, 1974); Eric Lott, *Love and Theft: Blackface Minstrelsy and the American Working Class* (New York: Oxford University Press, 1995); Michael Rogin, *Blackface, White Noise: Jewish Immigrants in the Hollywood Melting Pot* (Berkeley: University of California Press, 1996).

[5]Eric Foner, *Reconstruction: America's Unfinished Revolution, 1863–1877* (New York: Harper and Row, 1988).

[6]C. Vann Woodward, *Origins of the New South* (Baton Rouge: Louisiana State University Press, 1971).

[7]Stephen Jay Gould. *The Mismeasure of Man* (New York: Norton, 1996, rev. ed.).

[8]George M. Fredrickson, *The Black Image in the White Mind: The Debate on Afro-American Character and Destiny, 1817–1914* (New York: Harper and Row, 1971).

[9]Rayford W. Logan, *The Betrayal of the Negro: From Rutherford B. Hayes to Woodrow Wilson* (New York: Collier, 1965).

[10]W. E. B. Du Bois, *The Souls of Black Folk,* edited with an Introduction by David W. Blight and Robert Gooding-Williams (Boston: Bedford Books, 1997, originally published in 1903), 45; David Levering Lewis, *W. E. B. Du Bois: Biography of a Race, 1868–1919* (New York: Henry Holt, 1993); Louis R. Harlan, *Booker T. Washington: The Making of a Black Leader, 1856–1901* (New York: Oxford University Press, 1972).

[11]Louis R. Harlan, *Booker T. Washington: The Wizard of Tuskegee, 1901–1915* (New York: Oxford University Press, 1983), 244–51; August Meier, *Negro Thought in America, 1880–1915* (Ann Arbor: University of Michigan Press, 1963), 100–118; W. E. B. Du Bois, *The Autobiography of W. E. B. Du Bois* (New York: International Publishers, 1968), 236–76.

[12]Mark V. Tushnet, *The NAACP's Legal Strategy Against Segregated Education, 1925–1950* (Chapel Hill: University of North Carolina Press, 1987).

[13]Nathan Huggins, "The Deforming Mirror of Truth," in *Revelations: American History, American Myths* (New York: Oxford University Press, 1995), 252–83; Edmund Morgan, *American Slavery, American Freedom: The Ordeal of Colonial Virginia* (New York: Norton, 1975).

[14]Genna Rae McNeil. *Groundwork: Charles Hamilton Houston and the Struggle for Civil Rights* (Philadelphia: University of Pennsylvania Press, 1983); Judge A. Leon Higginbotham, Foreword, *Groundwork,* xv, xvii.

[15]McNeil, *Groundwork,* 82.

[16]Ibid., 111; Dan T. Carter, *Scottsboro* (New York: Oxford University Press, 1969); James Goodman, *Stories of Scottsboro* (New York: Pantheon Books, 1994).

[17]McNeil, *Groundwork,* 131–55, 223.

[18]Mary Frances Berry, *Black Resistance-White Law: A History of Constitutional Racism in America* (New York: A. Lane, Penguin Press, 1971 (1994 ed.), xi, xii.

[19]Houston cited in McNeil, *Groundwork,* 123.

[20]Ibid., 117–19, 76–85, 123, 216–18; Morton Horwitz, "The Jurisprudence of Brown and the Dilemmas of Liberalism," in Michael Namorato, ed., *Have We Overcome? Race Relations Since Brown* (Jackson: University Press of Mississippi, 1979).

[21]"Houstonian social engineering entailed five obligations for black lawyers," McNeil writes: "(1) to be 'prepared to anticipate, guide, and interpret group advancement'; . . . (2) to be the 'mouthpiece of the weak and a sentinel guarding against wrong'; . . . (3) to ensure that 'the course of change is . . . orderly with a minimum of human loss and suffering,' when possible 'guid[ing] . . . antagonistic and group forces into channels where they w[ould] not clash'; . . . (4) to recognize that the written constitution and inertia against its amendment give lawyers room for social experimentation and therefore, to 'use . . . the law as an instrument available to [the] minority unable to adopt direct action to achieve its place in the community and nation'; . . . (5) to engage in 'a carefully planned [program] to secure decisions, rulings and public opinion on . . . broad principle[s]' . . . while 'arousing and strengthening the local will to struggle.' " . . . McNeil, *Groundwork,* 217.

[22]Tushnet, *The NAACP's Legal Strategy,* 118.

[23]Ibid., 33.

[24]Ibid.

[25]Kluger, *Simple Justice,* 133–38; Tushnet, *The NAACP's Legal Strategy,* 16–17, 20–29.

[26]Kluger, *Simple Justice,* 136–37; Tushnet, *The NAACP's Legal Strategy,* 33–48.

[27]Houston cited in Tushnet, *The NAACP's Legal Strategy,* 34.

[28]Tushnet, *The NAACP's Legal Strategy,* 81, 88, 106–07; Kluger, *Simple Justice.*

[29]Berry, *Black Resistance/White Law;* Derrick Bell, *Race, Racism, and American Law,* 3d ed. (Boston: Little, Brown, 1992).

[30]Bluesman Carl Martin is cited in Lawrence W. Levine, "American Culture and the Great Depression," in *The Unpredictable Past: Explorations in American Cultural History* (New York: Oxford University Press, 1993).

[31]Jonathan Scott Holloway, "Confronting the Veil: New Deal African American Intellectuals and the Evolution of a Radical Voice," (Ph.D. dsst., Yale University, 1994); Nell I. Painter, *The Narrative of Hosea Hudson: His Life as a Negro Communist in the South* (Cambridge: Harvard University Press, 1979); *Robin D. G. Kelley, Hammer and Hoe: Alabama Communists during the Great Depression* (Chapel Hill: University of North Carolina Press, 1990).

[32]DuBois, *The Autobiography*, especially, 289–325.

[33]DuBois, W. E. B., "Does the Negro Need Separate Schools?" *The Journal of Negro Education*, IV:3 (July 1935), 328–35; Daryl Michael Scott, *Contempt and Pity: Social Policy and the Image of the Damaged Black Psyche, 1880–1996* (Chapel Hill: University of North Carolina Press, 1997).

[34]Du Bois, "Does the Negro Need Separate Schools?", 335. [Note: The excerpts shown here do not appear in the documents section of this book in order to avoid repetition.]

[35]Ibid., 331.

[36]Patricia Sullivan, *Days of Hope: Race and Democracy in the New Deal Era* (Chapel Hill: University of North Carolina Press, 1996), 141.

[37]Ibid., 142–43.

[38]Ibid.

[39]Gunnar Myrdal, *An American Dilemma: The Negro Problem and Modern Democracy* (New York: Harper and Brothers, 1944); Walter A. Jackson, *Gunnar Myrdal and America's Conscience: Social Engineering and Racial Liberalism, 1938–1987* (Chapel Hill: University of North Carolina Press, 1990); David Southern, *Gunnar Myrdal and Black-White Relations: The Use and Abuse of "An American Dilemma,"* (Baton Rouge: Louisiana State University Press, 1987).

[40]Myrdal, *An American Dilemma*; Jackson, *Gunnar Myrdal*.

[41]Tushnet, *The NAACP's Legal Strategy*, 105–37.

[42]Darlene Clark Hine, *Black Victory: The Rise and Fall of the White Primary in Texas* (Millwood, NY: KTO Press, 1979).

[43]See the documentary film "The Road to Brown: The Untold Story of 'The Man Who Killed Jim Crow.' " The second half of the film's title is a reference to the legendary civil rights lawyer Charles Hamilton Houston.

[44]Leonard W. Levy and Douglas L. Jones, *Jim Crow in Boston: The Origin of the Separate But Equal Doctrine* (New York: De Capo Press, 1974).

[45]James M. McPherson, *The Abolitionist Legacy* (Princeton: Princeton University Press, 1975).

[46]Brook Thomas, *Plessy v. Ferguson: A Brief History with Documents* (Boston: Bedford Books, 1997).

[47]For thoughtful discussions of race and the process of "racialization," see Michael Omi and Howard Winant, *Racial Formation in the United States: From the 1960s to the 1980s.* (New York: Routledge and Kegan Paul, 1986); Kimberle Crenshaw et al., eds., *Critical Race Theory: The Key Writings That Formed the Movement* (New York: New Press, 1995).

[48]Ibid.; William Wei, *The Asian Movement: A Social History* (Philadelphia: Temple University Press, 1993).

[49]Tushnet, *The NAACP Legal Strategy*, 70–137.

[50]Ibid., 161.

[51]McNeil, *Groundwork*, 176–85; Mark Tushnet and Katya Lezin, "What Really Happened in *Brown v. Board of Education*," *Columbia Law Review* 91:8 (December 1991), 1885.

[52]Attorney General Howard McGrath cited in Tushnet, *The NAACP Legal Strategy*, 130.

[53]Mark A. Chesler, Joseph Sanders, and Debra S. Kalmuss, *Social Science in Court: Mobilizing Experts in the School Desegregation Cases* (Madison: University of Wisconsin Press, 1988), 17–21.

[54]Scott, *Contempt and Pity*, 119–36.

[55]Chesler, Sanders, and Kalmuss, *Social Science in Court,* 17–25.

[56]Tushnet and Lezin, *"What Really Happened?",* 1890–91.

[57]Ibid., 1916–1917.

[58]Ibid., 1891, 1888.

[59]Ibid., 1922.

[60]Ibid., 1928.

[61]Tushnet, *The NAACP Legal Strategy,"* 143; Tushnet and Lezin, *"What Really Happened?",* 1884.

[62]Scott, *Contempt and Pity,* 130–36.

[63]Kluger, *Simple Justice,* 708, 710.

[64]Cleophus Brown cited in Shirley Moore, " 'To Place Our Deeds:' The African American Community in Richmond California" (Berkeley: University of California Press, forthcoming), ch. 5, p. 78; Robert E. Hemenway, *Zora Neale Hurston: A Literary Biography* (Urbana: University of Illinois Press, 1977), 329, 336. Hurston cited in Werner Sollors, "Of Mules and Mares in a Land of Difference; or, Quadrupeds All?," *American Quarterly,* 42 (June 1990), 171.

[65]Manning Marable, *W. E. B. Du Bois: Black Radical Democrat* (Boston: Twayne, 1986), 200.

[66]Harry M. Levin, "Education and Earnings of Blacks and the *Brown* Decision," in Michael Namorato, ed., *Have We Overcome?,* 110–11; Albert P. Blaustein and Clarence C. Ferguson, *Desegregation and the Law: The Meaning and Effect of the School Segregation Cases* (New Brunswick, NJ: Rutgers University Press, 1957), ix, 5.

[67]Tushnet, *"What Really Happened?",* 1867.

[68]For conservative accounts, see: Gerald N. Rosenberg, *The Hollow Hope: Can Courts Bring About Social Change?* (Chicago: University of Chicago Press, 1991); Michael Klarman, *"Brown,* Racial Change, and the Civil Rights Movement," *Virginia Law Review* 80:1 (February 1994), 7–150. For radical accounts, see: Derrick Bell, *"Brown and the Interest-Convergence Dilemma,"* in Bell, ed., *Shades of Brown: New Perspectives on School Desegregation* (New York: Teacher's College, Columbia University, 1980); Morton J. Horwitz, "The Jurisprudence of Brown and the Dilemmas of Liberalism, in Namorato, ed., *Have We Overcome,* 173–87.

[69]Horwitz, "The Jurisprudence of Brown," 186–87.

[70]Ibid., 185, 184.

[71]Ibid., 186.

[72]Ibid., 178.

[73]Lee Baker, "From Savage to Negro: Anthropology and the Construction of Race, 1896–1954," (Ph.D. dsst. Johns Hopkins, 1994).

[74]Scott, *Contempt and Pity,* xi.

[75]Martin Luther King, Jr., "The Civil Rights Struggle in the United States," *The Record of the Association of the Bar of the City of New York* 20 (1965), 5, 6, cited in Higginbotham's Foreword to McNeil, *Groundwork,* xv.

[76]"Soul Make A Path Through Shouting," in Cyrus Cassells, *Soul Make A Path Through Shouting* (Port Townsend, Washington: Copper Canyon Press, 1994), 18.

1

Roberts v. City of Boston (1849)

A Petition on Behalf of Black Inclusion in the Boston Common Schools

October 17, 1787

Historically, African Americans have understood the importance of public school education to the individual and collective elevation of their people. During the antebellum era, they shared the growing belief, especially widespread in "enlightened" northern communities such as Boston, Philadelphia, and New York City, in government's responsibility to maintain public schools open to the children of all citizens. Particularly within communities such as Boston, the evolving American success ethic, which blacks shaped and imbibed, stressed a common (public) school education as a key to solid citizenship as well as socioeconomic mobility. The following early petition from a group of black Bostonians to the Massachusetts legislature on behalf of a common school education for their children based its claims on the dual assumptions of citizenship and morality. Why do you think they argued their case on these grounds? What is your assessment of the substance and tone of the petition? Why do you think this effort failed?

To the Honorable Senate and House of Representatives of the Commonwealth of Massachusetts Bay, in General Court assembled.

The petition of a great number of blacks, freemen of this Commonwealth, humbly sheweth, that your petitioners are held in common with other freemen of this town and Commonwealth and have never been back-

Herbert Apetheker, *A Documentary History of the Negro People in the United States*, vol. 1 (New York: Citadel Press, 1961), 19–20.

ward in paying our proportionate part of the burdens under which they have, or may labor under; and as we are willing to pay our equal part of these burdens, we are of the humble opinion that we have the right to enjoy the privileges of free men. But that we do not will appear in many instances, and we beg leave to mention one out of many, and that is of the education of our children which now receive no benefit from the free schools in the town of Boston, which we think is a great grievance, as by woful experience we now feel the want of a common education. We, therefore, must fear for our rising offspring to see them in ignorance in a land of gospel light when there is provision made for them as well as others and yet can't enjoy them, and for not other reason can be given this they are black. . . .

We therefore pray your Honors that you would in your wisdom some provision may be made for the education of our dear children. And in duty bound shall ever pray.

MARIA W. STEWART
A Black Teacher's Travail
1850s

Public school education came to the South during the Reconstruction era (1863–77), in large measure because of the efforts of black legislators and black public pressure. Prior to the Civil War for most southerners — whites and free blacks alike — there were no public school systems. Indeed, as Roberts v. Boston *demonstrated, antiblack prejudice all too often kept free blacks out of public schools even in the North. Churches, family members, self-improvement organizations, subscription or "pay" schools (private schools for the working class and the poor), and other tutorial arrangements from the formal to the ad hoc provided many blacks with the rudiments of education. As this document shows, these efforts were often quite difficult to sustain over time.*

Maria W. Stewart, "Sufferings during the War," in *Maria W. Stewart: America's First Black Woman Political Writer — Essays and Speeches* ed. Marilyn Richardson (Bloomington: Indiana University Press, 1987), 98–99.

Maria W. Stewart (1803–1879) was a deeply religious black woman highly committed to the elevation of blacks, women as well as men. A speaker, writer, and teacher, she experienced serious problems attempting to keep afloat a school for black children in the early 1850s in Baltimore, a major port city with a sizable black population in an Upper South slave state. During the early 1830s in Boston, she had become the first American-born woman, white or black, to lecture publicly. The notoriety she created among blacks as well as whites for leaving woman's domestic sphere and transgressing the male public sphere forced her to abandon what appeared to be a promising public speaking career. This document details her valiant (yet ultimately unsuccessful) efforts at running a school. Why do you think she turned to education and teaching once her career as a public speaker was thwarted? What role did religion play in this endeavor? What can we gather about the quest for black education in the antebellum South from this episode?

Having lost my position at Williamsburg, Long Island, and hearing the colored people were more religious and God-fearing in the South, I wended my way to Baltimore in 1852. But I found all was not gold that glistened; and when I saw the want of means for the advancement of the common English branches, with no literary resources for the improvement of the mind scarcely, I threw myself at the foot of the Cross, resolving to make the best of a bad bargain. And there I lay; and then arose, in the strength of the Lord and in the power of His might, wrote my programme, printed and issued my circulars stating I would open school and would teach reading, writing, spelling, mental and practical arithmetic, and whatever other studies called for.

Not knowing the prices, I found myself teaching every branch for 50 cents per month, until informed by another teacher that no writing was taught for less than $1 per month. Bought wit is the dearest wit. I have never been very shrewd in money matters; and being classed as a lady among my race all my life, and never exposed to any hardship, I did not know how to manage. I had been teaching in New York and Williamsburg, and had the means of always paying my way. But when I came to teach a pay school I found the difference.

But God promised that my bread and water should be sure; and having food and raiment I was content. I would make enough just to supply my wants for the time being, but not a dollar over. I did not make any charge for wood and coal. And always had that refined sentiment of delicacy about me that I could not bear to charge for the worth of my labor. If any loss was to be sustained the loss was always on my side, and not on the side of the parent or the scholar.

Fugitive Slave Poster
1851

Northern free blacks saw their struggles as inextricably bound to those of southern blacks, slave and free. The Fugitive Slave Act of 1850, like previous statutes empowering slave owners and hired slave catchers to pursue their runaway property even on free soil in the North, both represented and tightened that bond. The law also further alienated northern free blacks from the federal government. While often severely circumscribed, freedom for blacks in the North was more extensive than that for free blacks in the South. This relatively freer environment sustained a vigorous public tradition of protest and vigilance among blacks in the North. Consequently, the Fugitive Slave Act engendered widespread protest among northern blacks and their allies.

Like Chief Justice Roger Taney's ruling in Scott v. Sandford *(1857), which nullified the claims of free blacks to national citizenship, the Fugitive Slave Law of 1850 clearly signaled a further decline in the fortunes of free blacks and intensified the worries of northern blacks about their legal rights and the diminishing quality of their lives. As a result, the 1850s saw an increase in black nationalist and black separatist developments, including efforts to encourage black emigration to points outside the United States, like West Africa and Haiti.*

A particularly dreaded consequence of the Fugitive Slave Act of 1850 was the increased numbers of free as well as fugitive slave blacks kidnapped and enslaved under its aegis. This profound threat to northern black freedom only exacerbated the outcry about the injustice of the statute and its prosecution. Many northern whites were dismayed by the law's serious threat to black personal liberty and the flagrant abuses of the slave-catching network. Intense protest led by blacks and their allies resulted in a series of personal liberty laws seeking to protect the besieged freedom of northern blacks.

The poster reprinted here is a response to the Boston imprisonment of a Virginia fugitive, Shadrach, and the successful effort of Boston's African

Prints and Photographs Division, Library of Congress.

CAUTION!!

COLORED PEOPLE

OF BOSTON, ONE & ALL,

You are hereby respectfully CAUTIONED and advised, to avoid conversing with the

Watchmen and Police Officers of Boston,

For since the recent ORDER OF THE MAYOR & ALDERMEN, they are empowered to act as

KIDNAPPERS

AND

Slave Catchers,

And they have already been actually employed in KIDNAPPING, CATCHING, AND KEEPING SLAVES. Therefore, if you value your LIBERTY, and the *Welfare of the Fugitives* among you, *Shun* them in every possible manner, as so many *HOUNDS* on the track of the most unfortunate of your race.

Keep a Sharp Look Out for KIDNAPPERS, and have TOP EYE open.

APRIL 24, 1851.

Americans to free him. Political posters such as this were a quite common and vital visual mode of urban political discourse, notably for the organization of groups and protests as well as for the dissemination of political statements. Why do you think Boston's black community chose to express this particular concern in this way? What does the poster suggest about the possible roles of this often ephemeral form of communication in early black battles for justice?

CHARLES SUMNER

Brief for Public School Integration

1849

Benjamin Roberts, a black activist in the 1840s struggle to integrate Boston's public (common) schools, initiated a suit on behalf of his five-year-old daughter Sarah Roberts. The suit, Roberts v. City of Boston, *challenged the city's practice of racially segregating its schools, a practice that paradoxically had been initiated by blacks themselves.*

In the early 1800s, Boston's blacks had sought a separate black school as a way not only to provide education for their children but also to shield them from the virulent race prejudice within the public schools. In 1818 the Boston School Committee took control of the private and separate all-black school. But by the 1840s, the poor quality of education in the school led to conflicting ideas among blacks about how best to improve it. Some argued for a better white teacher. Others contended that a black teacher would be better equipped to deal with the demands of black students. An apparent majority, however, favored abolishing the segregated school and establishing fully integrated schools. Benjamin Roberts was part of an active group of blacks supporting this cause, along with white abolitionists such as Wendell Phillips and William Lloyd Garrison. Having successfully battled railroad car segregation in the early 1840s, by the end of the decade they were focusing much of their protest energies on school desegregation.

Only 2 percent of Boston's population, blacks were confined to one of the city's roughly 117 primary schools in the mid-1840s. Even in the most liberal parts of the antebellum North, like Boston, free blacks endured pervasive antiblack prejudice and discrimination. Sarah's application to attend a white school had been rejected four times. For the plaintiff's supporters, the fact that she passed five white schools on her way to the black one only exacerbated the injustice. The plaintiff's suit, argued principally by the eminent lawyer, abolitionist, and future Massachusetts senator Charles Sumner (1811–1874), charged that this invidious racial discrimination violated Sarah's right to equality before the law, or what he called "precise equality." Sumner's able co-counsel was Robert E. Davis, one of only two black lawyers in the state.

Charles Sumner, "Equality before the Law: Unconstitutionality of Separate Colored Schools in Massachusetts," in *Charles Sumner: His Complete Works* (New York: Negro Universities Press, 1969), 52–54, 64–66, 68–69, 70–76, 80–100.

The Boston School Committee argued that the racially based discrimination was legal; it did not deny the plaintiff her legal right to an education. Rather, it merely reflected a view of the law as narrow, pragmatic, and consistent with white supremacy. In addition, the School Committee presented this exercise of local discretionary power as fully within its purview.

A formal and precise moral argument, Sumner's brief in support of the plaintiff in Roberts v. Boston *is a grand and eloquent statement on behalf of human rights as its guiding legal principle. He argues forcefully for a tradition of equality before the law emanating from the Massachusetts state constitution as well as the Declaration of Independence. Understanding that he speaks to a much larger audience, he uses the brief as a platform to articulate the Enlightenment-inspired causes of equality and abolitionism. The brief's radical espousal of these causes was bold and utopian. While the state court's 1849 ruling in* Roberts v. Boston *went against Sumner's position, in 1855 the Massachusetts legislature accepted his argument by passing a statute outlawing segregated schools. That Sumner's brief has inspired and contributed significantly to civil rights law, especially as enunciated in the* Brown *case, speaks volumes for its prescience and power.*

Do you agree that Sumner's brief demonstrates "prescience and power"? To answer that question, you might consider a series of related questions. What does the brief tell us about the state of race relations in the antebellum United States? How does Sumner define law, equality, and race? How does he present the patterns of interaction among these concepts? What is his view of the relationship between law and society?

. . . Can any discrimination on account of race or color be made among children entitled to the benefit of our Common Schools under the Constitution and Laws of Massachusetts? This is the question which the Court is now to hear. . . .

Or, stating the question with more detail, and with more particular application to the facts of the present case, are the Committee having superintendence of the Common Schools of Boston intrusted with *power,* under the Constitution and Laws of Massachusetts, to exclude colored children from the schools, and compel them to find education at separate schools, set apart for colored children only, at distances from their homes less convenient than schools open to white children?

This important question arises in an action by a colored child only five years old, who, *by her next friend,* sues the city of Boston for damages on account of a refusal to receive her into one of the Common Schools.

It would be difficult to imagine any case appealing more strongly to your best judgment, whether you regard the parties or the subject. On the one side is the City of Boston, strong in wealth, influence, character; on the other side is a little child, of degraded color, of humble parents, and still within the period of natural infancy, but strong from her very weakness, and from the irrepressible sympathies of good men, which, by a divine compensation, come to succor the weak. This little child asks at your hands her *personal rights*. So doing, she calls upon you to decide a question which concerns the personal rights of other colored children, — which concerns the Constitution and Laws of the Commonwealth, — which concerns that *peculiar institution* of New England, the Common Schools, — which concerns the fundamental principles of human rights, — which concerns the Christian character of this community. Such parties and such interests justly challenge your earnest attention. . . .

The Question Stated

. . . Of Equality I shall speak, not only as a sentiment, but as a principle embodied in the Constitution of Massachusetts, and obligatory upon court and citizen. It will be my duty to show that this principle, after finding its way into our State Constitution, was recognized in legislation and judicial decisions. Considering next the circumstances of this case, it will be easy to show how completely they violate Constitution, legislation, and judicial proceedings,—*first,* by subjecting colored children to inconvenience inconsistent with the requirements of Equality, and, *secondly,* by establishing a system of Caste odious as that of the Hindoos, — leading to the conclusion that the School Committee have no such power as they have exercised, and that it is the duty of the Court to set aside their unjust by-law. In the course of this discussion I shall exhibit the true idea of our Common Schools, and the fallacy of the pretension that any exclusion or discrimination founded on race or color can be consistent with Equal Rights. . . .

Equality Under Constitution of
Massachusetts and Declaration of Independence

. . . The Declaration of Independence, which followed the French Encyclopedia[1] and the political writings of Rousseau,[2] announces among self-evident truths, *"that all men are created equal;* that they are endowed by their Cre-

[1]Denis Diderot's collection of writings by French intellectuals of the Enlightenment (18th c. Europe) was published from 1751 to 1772.
[2]Jean-Jacques Rousseau, Swiss-French writer and major thinker of the Enlightenment (1712–1778).

ator with certain unalienable rights; that among these are life, liberty, and the pursuit of happiness." The Constitution of Massachusetts repeats the same truth in a different form, saying, in its first article: *"All men are born free and equal,* and have certain natural essential, and unalienable rights, among which may be reckoned the right of enjoying and defending their lives and liberties." Another article explains what is meant by Equality, saying: "No man, nor corporation or association of men, have any other title to obtain advantages, or particular and exclusive privileges, distinct from those of the community, than what arises from the consideration of services rendered to the public. . . ." This language, in its natural signification, condemns every form of inequality in civil and political institutions.

. . . And permit me to say, it is a childish sophism to adduce any physical or mental inequality in argument against Equality of Rights.

Obviously, men are not born equal in physical strength or in mental capacity, in beauty of form or health of body. Diversity or inequality in these respects is the law of creation. From this difference springs divine harmony. But this inequality is in no particular inconsistent with complete civil and political equality.

The equality declared by our fathers in 1776, and made the fundamental law of Massachusetts in 1780, was *Equality before the Law.* Its object was to efface all political or civil distinctions, and to abolish all institutions founded upon *birth.* . . . Here is the Great Charter of every human being drawing vital breath upon this soil, whatever may be his condition, and whoever may be his parents. He may be poor, weak, humble, or black, — he may be of Caucasian, Jewish, Indian, or Ethiopian race, — he may be of French, German, English, or Irish extraction; but before the Constitution of Massachusetts all these distinctions disappear. He is . . . a Man, the equal of all his fellow-men. He is one of the children of the State, which, like an impartial parent, regards all its offspring with an equal care. To some it may just allot higher duties, according to higher capacities; but it welcomes all to its equal hospitable board. The State, imitating the divine justice, is no respecter of persons. . . .

Separate Schools Inconsistent with Equality

It is easy to see that the exclusion of colored children from the Public Schools is a constant inconvenience to them and their parents, which white children and white parents are not obliged to bear. Here the facts are plain and unanswerable, showing a palpable violation of Equality. *The black and white are not equal before the law.* I am at a loss to understand how anybody can assert that they are.

Among the regulations of the Primary School Committee is one to this effect. "Scholars to go to the school nearest their residences. Applicants for admission to our schools (with the exception and provision referred to in the preceding rule) are especially entitled to enter the schools nearest to their places of residence." The exception here is "of those for whom special provision has been made" in separate schools, — that is, colored children.

In this rule — without the unfortunate exception — is part of the beauty so conspicuous in our Common Schools. . . . It may also be the boast of our Common Schools, that, through the multitude of schools, education in Boston is brought to every *white* man's door. But it is not brought to every *black* man's door. He is obliged to go for it, to travel for it, to walk for it, — often a great distance. . . .

Such a fact is sufficient to determine this case. If it be met by the suggestion, that the inconvenience is trivial, and such as the law will not notice, I reply, that it is precisely such as to reveal an existing inequality, and therefore the law cannot fail to notice it. There is a maxim . . . that even a trivial fact may give occasion to an important application of the law. . . . Also from the best examples of our history we learn that the insignificance of a fact cannot obscure the grandeur of the principle at stake. It was a paltry tax on tea, laid by a Parliament where they were not represented, that aroused our fathers to the struggles of the Revolution. They did not feel the inconvenience of the tax, but they felt its oppression. They went to war for a principle. Let it not be said, then, that in the present case the inconvenience is too slight to justify the appeal I make in behalf of colored children for *Equality before the Law*.

Looking beyond the facts of this case, it is apparent that the inconvenience from the exclusion of colored children is such as to affect seriously the comfort and condition of the African race in Boston. The two Primary Schools open to them are in Belknap Street and Sun Court. I need not add that the whole city is dotted with schools open to white children. Colored parents, anxious for the education of their children, are compelled to live in the neighborhood of the schools, to gather about them. . . . The liberty which belongs to the white man, of choosing his home, is not theirs. . . .

Separate Schools Are in the Nature of Caste

The separation of children in the Schools, on account of race or color, is in the nature of *Caste,* and, on this account, a violation of Equality. The case shows expressly that the child was excluded from the school nearest to her dwelling — the number in the school at the time warranting her admission—"on the sole ground of color." The first Majority Report presented to the School Committee, and mentioned in the statement of facts, presents the grounds of this discrimination with more fullness, saying, "It is one of *races,* not of *colors* merely. The distinction is one which the Allwise Creator has seen fit to establish; and it is founded deep in the physical, mental, and moral natures of the two races. No legislation, no social customs, can efface this distinction."[3] Words cannot be chosen more apt than these to describe the heathenish relation of Caste. . . .

School Committee Have No Power to Discriminate on Account of Color

. . . In entire harmony with the Constitution, the law says expressly what the Committee shall do. . . . The power to determine the "number" is easily executed, and admits of no question. The power to determine the "qualifications," though less simple, must be restricted to age, sex, and fitness, moral and intellectual. The fact that a child is black, or that he is white, cannot of itself be a qualification or a disqualification. Not to the skin can we look for the criterion of fitness.

It is sometimes pretended, that the Committee, in the exercise of their power, are intrusted with a discretion, under which they may distribute, assign, and classify all children belonging to the schools *according to their best judgment,* making, if they think proper, a discrimination of race or color. Without questioning that they are intrusted with a discretion, it is outrageous to suppose that their discretion can go to this extent. The Committee can have no discretion which is not in harmony with the Constitution and Laws. Surely they cannot, in any mere discretion, nullify a sacred and dear-bought principle of Human Rights expressly guarantied by the Constitution.

Regulations of Committee Must Be Reasonable

Still further,—and here I approach a more technical view of the subject,—it is an admitted principle, that the regulations and by-laws of municipal corporations must be *reasonable,* or they are inoperative and void. . . .

Assuming that this principle is applicable to the School Committee, their regulations and by-laws must be *reasonable.* Their discretion must be exercised in a reasonable manner. And this is not what the Commit-

[3] Report to the Primary School Committee, June 15, 1846, on the Petition of Sundry Colored Persons for the Abolition of the Schools for Colored Children, p. 7.

tee or any other body of men think reasonable, but what is reasonable in the eye of the Law. It must be *legally reasonable*. It must be approved by the *reason* of the Law. . . .

It is clear that the Committee may classify scholars according to age and sex, for the obvious reasons that these distinctions are inoffensive, and that they are especially recognized as *legal* in the law relating to schools. They may also classify scholars according to moral and intellectual qualifications, because such a power is necessary to the government of schools. But the Committee cannot assume, *a priori*, and without individual examination, that all of an *entire race* are so deficient in proper moral and intellectual qualifications as to justify their universal degradation to a class by themselves. Such an exercise of discretion must be unreasonable, and therefore illegal.

Separate School Not an Equivalent for Common School

But it is said that the School Committee, in thus classifying the children have not violated any principle of Equality, inasmuch as they provide a school with competent instructors for colored children, where they have advantages equal to those provided for white children. It is argued, that, in excluding colored children from Common Schools open to white children, the Committee furnish an *equivalent*.

Here there are several answers. I shall touch them briefly, as they are included in what has been already said.

1. The separate school for colored children is not one of the schools established by the law relating to Public Schools. It is not a Common School. As such it has no legal existence, and therefore cannot be a *legal equivalent*. . . .

2. The second is that in point of fact the separate school is not an equivalent. We have already seen that it is the occasion of inconvenience to colored children, which would not arise, if they had access to the nearest Common School, besides compelling parents to pay an additional tax, and inflicting upon child and parent the stigma of Caste. Still further, — and this consideration cannot be neglected, — the matters taught in the two schools may be precisely the same, but a school exclusively devoted to one class must differ essentially in spirit and character from that Common School known to the law, where all classes meet together in Equality. It is a mockery to call it an equivalent.

3. But there is yet another answer. Admitting that it is an equivalent, still the colored children cannot be compelled to take it. Their rights are found in Equality before the Law; nor can they be called to renounce one

jot of this. They have an equal right with white children to the Common Schools. A separate school, though well endowed, would not secure to them that precise Equality which they would enjoy in the Common Schools. The Jews in Rome are confined to a particular district called the Ghetto, and in Frankfort[4] to a district known as the Jewish Quarter. It is possible that their accommodations are as good as they would be able to occupy, if left free to choose throughout Rome and Frankfort; but this compulsory segregation from the mass of citizens is of itself an *inequality* which we condemn. It is a vestige of ancient intolerance directed against a despised people. It is of the same character with the separate schools in Boston. . . .

Disastrous Consequences of Power to Make Separate Schools

In determining that the School Committee have no *power* to make this discrimination we are strengthened by another consideration. If the power exists in the present case, it cannot be restricted to this. The Committee may distribute all the children into classes, according to mere discretion. They may establish a separate school for Irish or Germans, where each may nurse an exclusive nationality alien to our institutions. They may separate Catholics from Protestants, or, pursuing their discretion still further, may separate different sects of Protestants, and establish one school for Unitarians, another for Methodists. They may establish a separate school for the rich, that the delicate taste of this favored class may not be offended by the humble garments of the poor. They may exclude the children of mechanics, and send them to separate schools. All this, and much more, can be done in the exercise of that high-handed power which makes a discrimination on account of race or color. The grand fabric of our Common Schools, the pride of Massachusetts, — where, at the feet of the teacher, innocent childhood should come, unconscious of all distinctions of birth, — where the Equality of the Constitution and of Christianity should be inculcated by constant precept and example, — will be converted into a heathen system of proscription and Caste. We shall then have many different schools, representatives of as many different classes, opinions, and prejudices; but we shall look in vain for the true Common School of Massachusetts. Let it not be said that there is little danger that any Committee will exercise a discretion to this extent. They must not be intrusted with the power. Here is the only safety worthy of a free people. . . .

[4] Frankfurt, Germany.

Origin of Separate Schools

In extenuation of the Boston system, it is sometimes said that the separation of white and black children was originally made at the request of colored parents. This is substantially true. . . . Much I fear that the inhuman bigotry of Caste—sad relic of the servitude from which they had just escaped—was at this time too strong to allow colored children kindly welcome in the free schools, and that, from timidity and ignorance, they hesitated to take a place on the same benches with the white children. Perhaps the prejudice was so inveterate that they could not venture to assert their rights. In 1800 a petition from sixty-six colored persons was presented to the School Committee, requesting the establishment of a school for their benefit. Some time later, private munificence came to the aid of this work, and the present system of separate schools was brought into being.

These are interesting incidents belonging to the history of the Boston Schools, but they cannot in any way affect the rights of colored people or the powers of the School Committee. These rights and these powers stand on the Constitution and Laws. . . . [A]ll must agree the assent of a few to an unconstitutional and illegal course nearly half a century ago, when their rights were imperfectly understood, cannot alter the Constitution and the Laws so as to bind their descendants forever in the thrall of Caste. Nor can the Committee derive from this assent, or from any lapse of time, powers in derogation of the Constitution and the Rights of Man.

It is clear that the sentiments of the colored people have now changed. The present case, and the deep interest which they manifest in it, thronging the Court to watch this discussion, attest the change. With increasing knowledge they have learned to know their rights, and feel the degradation to which they are doomed. . . . Their present effort is the token of a manly character, which this Court will respect and cherish.

Evils of Separate Schools

But it is said that these separate schools are for the benefit of both colors, and of the Public Schools. In similar spirit Slavery is sometimes said to be for the benefit of master and slave, and of the country where it exists. There is a mistake in the one case as great as in the other. This is clear. Nothing unjust, nothing ungenerous, can be for the benefit of any person or any thing. From some seeming selfish superiority, or from the gratified vanity of class, short-sighted mortals may hope to draw permanent good; but even-handed justice rebukes these efforts and redresses the wrong.

The whites themselves are injured by the separation. Who can doubt this? With the Law as their monitor, they are taught to regard a portion of the human family, children of God, created in his image, coequals in his love, as a separate and degraded class; they are taught practically to deny that grand revelation of Christianity, the Brotherhood of Man. Hearts, while yet tender with childhood, are hardened, and ever afterward testify to this legalized uncharitableness. Nursed in the sentiments of Caste, receiving it with the earliest food of knowledge, they are unable to eradicate it from their natures, and then weakly and impiously charge upon our Heavenly Father the prejudice derived from an unchristian school. Their characters are debased, and they become less fit for the duties of citizenship. . . .

Who can say that this does not injure the blacks? Theirs, in its best estate, is an unhappy lot. A despised class, blasted by prejudice and shut out from various opportunities, they feel this proscription from the Common Schools as a peculiar brand. Beyond this, it deprives them of those healthful, animating influences which would come from participation in the studies of their white brethren. It adds to their discouragements. It widens their separation from the community, and postpones that great day of reconciliation which is yet to come. . . .

. . . A degraded or neglected class, if left to themselves, will become more degraded or neglected. . . . Happily, our educational system, by the blending of all classes, draws upon the whole school that attention which is too generally accorded only to the favored few, and thus secures to the poor their portion of the fruitful sunshine. But the colored children, placed apart in separate schools, are deprived of this peculiar advantage. Nothing is more clear than that the welfare of classes, as well as of individuals, is promoted by mutual acquaintance. Prejudice is the child of ignorance. It is sure to prevail, where people do not know each other. Society and intercourse are means established by Providence for human improvement. They remove antipathies, promote mutual adaptation and conciliation, and establish relations of reciprocal regard. . . .

Duty of the Court

. . . Still, which way soever we turn we are brought back to one single proposition,—*the Equality of men before the Law.* This stands as the mighty guardian of the colored children in this case. It is the constant, ever-present, tutelary genius of this Commonwealth, frowning upon every privilege of birth, every distinction of race, every institution of Caste. You cannot slight it or avoid it. You cannot restrain it. God grant that you may welcome it! Do this, and your words will be a "charter and freehold of rejoicing" to a race which by much suffering has earned a title to much regard. Your judgment will become a sacred landmark, not in jurisprudence only, but in the history of Freedom, giving precious encouragement to the weary and heavy-laden wayfarers in this great cause. Massachusetts, through you, will have fresh title to respect, and be once more, as in times past, an example to the whole land. . . . In the Caste Schools of Boston the prejudice of color seeks its final refuge. It is for you to drive it forth. You do well, when you rebuke and correct individual offences; but it is a higher office to rebuke and correct a vicious institution. Each individual is limited in influence; but an institution has the influence of numbers organized by law. The charity of one man may counteract or remedy the uncharitableness of another; but no individual can counteract or remedy the uncharitableness of an organized injury. Against it private benevolence is powerless. It is a monster to be hunted down by the public and the constituted authorities. And such is the institution of Caste in the Common Schools of Boston, which now awaits a just condemnation from a just Court. . . .

Slavery, in one of its enormities, is now before you for judgment. Hesitate not, I pray you, to strike it down. Let the blow fall which shall end its domination here in Massachusetts. . . .

MASSACHUSETTS CHIEF JUSTICE
LEMUEL SHAW

Opinion of the Court in Roberts v. City of Boston

1849

The distinguished jurist Chief Justice Lemuel Shaw and his colleagues disagreed with Sumner's brief. Siding with the Boston School Committee's argument instead, Shaw, speaking for the state's highest court, found the practice of a segregated school for blacks to be a legal exercise of the School

As reprinted in Harold W. Horowitz and Kenneth L. Karst, eds., *Law, Lawyers and Social Change: Cases and Materials on the Abolition of Slavery, Racial Segregation, and Inequality of Educational Opportunity* (Indianapolis: Bobbs-Merrill, 1969), 117–19.

Committee's authority, fully consistent with local custom. Because Shaw was an acknowledged constitutional expert on the states' use of their broad police powers for the common good, his argument that this policy was a reasonable exercise of power was significant. He also emphasized the distinction between the ideal and the reality of equality as expressed in civil rights. Law, he stressed, was built upon the tradition of actual equality. In that tradition, the state had the power to define the specific nature and regulation of civil rights. He thus found the provision of a separate and substantially equal school for blacks to be consistent with the state constitution's recognition of the plaintiff's right to equality before the law.

Six years later, however, the Massachusetts legislature outlawed segregated schools. Unfortunately, the die was cast and this important legislative triumph for equal school rights would be vastly overshadowed by the wide-ranging impact of Shaw's judicial decision favoring separate schools.

This Massachusetts Supreme Judicial Court ruling became a critical element in the powerful legal edifice of Jim Crow. As a result, it has had a profound impact on race relations in American society since the late nineteenth century. Specifically, Shaw's ruling has typically been cited as compelling legal authority in a broad array of U.S. Supreme Court and lower court decisions supporting white supremacy, as in Plessy v. Ferguson *(1896), the most authoritative Supreme Court ruling in support of Jim Crow.*

In effect, the extraordinary influence of Shaw's decision owed principally to its complicity in the legal fiction of separate but equal, not the strength of its legal arguments. Do you agree? Why or why not? In this regard, it would be useful to work through several other questions, some of which relate also to Sumner's brief (p. 47). What does Shaw's decision tell us about the state of race relations in the antebellum United States? What are the arguments underlying his overall thesis? How does he define law and equality? What are the ramifications of these definitions for blacks both as a race and as citizens? How does Shaw interpret the relationship between law and society? Overall would you assess the decision's rhetoric or language as effective? Why or why not?

. . . The great principle, advanced by the learned and eloquent advocate of the plaintiff, is, that by the constitution and laws of Massachusetts, all persons without distinction of age or sex, birth or color, origin or condition, are equal before the law. This, as a broad general principle, such as ought to appear in a declaration of rights, is perfectly sound; it is not only expressed in terms, but pervades and animates the whole spirit of our constitution of free government. But, when this great principle comes to be applied to the actual and various conditions of persons in society, it will not warrant the assertion, that men and women are legally clothed with the same civil and political powers, and that children and adults are legally to have the same functions and be subject to the same treatment; but only that the rights of all, as they are settled and regulated by law, are equally entitled to the paternal consideration and protection of the law, for their maintenance and security. What those rights are, to which individuals, in the infinite variety of circumstances by which they are surrounded in society, are entitled, must depend on laws adapted to their respective relations and conditions.

Conceding, therefore, in the fullest manner, that colored persons, the descendants of Africans, are entitled by law, in this commonwealth, to equal rights, constitutional and political, civil and social, the question then arises, whether the regulation in question, which provides separate schools for colored children, is a violation of any of these rights.

Legal rights must, after all, depend upon the provisions of law; certainly all those rights of individuals which can be asserted and maintained in any judicial tribunal. The proper province of a declaration of rights and constitution of government, after directing its form, regulating its organization and the distribution of its powers, is to declare great principles and fundamental truths, to influence and direct the judgment and conscience of legislators in making laws, rather than to limit and control them, by directing what precise laws they shall make. The provision, that it shall be the duty of legislatures and magistrates to cherish the interests of literature and the sciences, especially the university at Cambridge, public schools, and grammar schools, in the towns, is precisely of this character. Had the legislature failed to comply with this injunction, and neglected to provide public schools in the towns, or should they so far fail in their duty as to repeal all laws on the subject, and leave all education to depend on private means, strong and explicit as the direction of the constitution is, it would afford no remedy or redress to the thousands of the rising generation, who now depend on these schools to afford them a most valuable education, and in introduction to useful life. . . .

In the absence of special legislation on this subject, the law has vested the power in the committee to regulate the system of distribution and classification; and when this power is reasonably exercised, without being abused or perverted by colorable pretences, the decision of the committee must be deemed conclusive. The committee, apparently upon great deliberation, have come to the conclusion, that the good of both classes of schools will be best promoted, by maintaining the separate primary schools for colored and for white children, and we can perceive no ground to doubt, that this is the honest result of their experience and judgment.

It is urged, that this maintenance of separate schools tends to deepen and perpetuate the odious distinction of caste, founded in a deep-rooted prejudice in public opinion. This prejudice, if it exists, is not created by law, and probably cannot be changed by law. Whether this distinction and prejudice, existing in the opinion and feelings of the community, would not be as effectually fostered by compelling colored and white children to associate together in the same schools, may well be doubted; at all events, it is a fair and proper question for the committee to consider and decide upon, having in view the best interest of both classes of children placed under their superintendence, and we cannot say, that their decision upon it is not founded on just grounds of reason and experience, and in the results of a discriminating and honest judgment. The increased distance, to which the plaintiff was obliged to go to school from her father's house, is not such, in our opinion, as to render the regulation in question unreasonable, still less illegal.

On the whole the court are of opinion, that upon the facts stated, the action cannot be maintained.

2

Plessy v. Ferguson (1896)

HENRY McNEAL TURNER

"Civil Rights: The Outrage of the Supreme Court of the United States upon the Black Man"

1889

Slaves achieved their freedom in the Civil War years through their own efforts, such as serving in the Union army or absconding to the Union lines, as well as through wartime legal measures such as Abraham Lincoln's Emancipation Proclamation (1863). During Reconstruction (1863–77), the government moved to make ex-slaves full-fledged Americans by abolishing slavery officially (Thirteenth Amendment, 1865), defining blacks as citizens (Fourteenth Amendment, 1868), and removing race as a barrier to the vote (Fifteenth Amendment, 1870). Constitutional law now extended civil rights to blacks and gave them the tools to sustain those rights. The Civil Rights Law of 1875 furthered black civil rights by banning racial discrimination in public accommodations, transportation, theaters, and civil arenas such as jury service. This federal statute clarified and expanded the meanings of civil rights and citizenship as outlined in the Fourteenth Amendment.

The flip side of the constitutional protection of blacks' civil rights was the racist backlash against blacks during Reconstruction and an orgy of murderous violence in its aftermath. This rapidly escalating white supremacism was buttressed by the federal government's abandonment of blacks after Reconstruction, exemplified by the growing number of antiblack court rulings. The Supreme Court decision in the Civil Rights Cases *(1883) —five*

Herbert Aptheker, *A Documentary History of the Negro People in the United States,* vol. 1 (New York: Citadel Press, 1961), 171–78.

separate challenges to the Civil Rights Law of 1875 that the Court heard together—signified that betrayal.

The Court nullified the 1875 law, ruling that it went far beyond the Fourteenth Amendment, which forbade racial discrimination by the state, not by private parties. In other words, the 1875 Civil Rights Law transgressed the boundary of private rights and social rights. This narrowing of the civil rights protections and remedies available to blacks deeply angered them and their allies. Nowhere was this rage better expressed than in the fiery rhetoric of Henry M. Turner (1834–1915), a bishop in the African Methodist Episcopal Church, a black nationalist and a promoter of a black back-to-Africa project in the 1890s. The following excerpt from a letter to a newspaper editor detailed Turner's denunciation of that "disgraceful" decision. Note that Turner praises Justice John Harlan's dissenting opinion upholding the legality of the 1875 Civil Rights Law. Indeed, that important dissent prefigured Harlan's even more influential dissent in Plessy v. Ferguson. How do you explain the depth of Turner's dismay? How does his religiosity figure in his arguments? What connection does Turner draw between civil rights on the one hand and black success (notably those who have achieved middle-class respectability) on the other? Do you find Turner persuasive and his rhetorical style effective? Why or why not?

Editor of the New York Voice

Amidst multitudinous duties I find, calling my attention, your note of recent date, asking me to briefly refer to the "Civil Rights Decisions," which, since their delivery has drawn from me expressions which many are pleased to call severe adverse strictures upon the highest court in this country, and upon all of its judges save one, Mr. Justice Harlan. It is to me a matter of that kind of surprise called wonder suddenly excited to find a single, solitary individual who belongs in the United States, or who has been here for any considerable time, unacquainted with those famous FIVE DEATH DEALING DECISIONS. Indeed, sir, those decisions have had, since the 15th day of October, A.D. 1883, the day of their pronouncement, more of my study than any other civil subject. I incline to the opinion that I have an argument which, taken as a concomitant of the learned dissenting sentiments of that eminent jurist, Mr. Justice Harlan, would to a rational mind make the judgment of Justice Bradley[1] and his associates a deliquescence—a bubble on the wave of equity—a legal nothing. You bid me in my reply to observe brevity. Shortness and conciseness seem

---

[1] Justice Joseph P. Bradley, author of the majority opinion.

to be the ever present rule when the Negro and his case is under treatment. However, I am satisfied that in saying this, I do not convey your reason for commanding me to condense, "boil down." The more I ponder the non-agreeing words of that member of our chief assize, who had the moral courage to bid defiance to race prejudice, the more certain am I that no words of mine, condemnatory of that decision, have been sufficiently harsh.

March 1st, 1875, Congress passed an act entitled "An act for the prevention of discrimination on the ground of race, color or previous condition of servitude," said act being generally known as the Civil Rights Bill, introduced during the lifetime of the Negro's champion, the immortal Charles Sumner. . . .

. . . The questions that come forward and will not down are: Was this law just? Did this law violate the principle which should be foremost in every hall of legislation—hurt no one, give unto every man his just due? Should the color of one's skin deny him privileges any more than the color of one's hair, seeing that the individual had nothing to do with the cause for the one or for the other? Before attempting to answer the above questions, which must and will suggest themselves to every *compos mentis*[2] we state the constitutional amendments upon which the act under consideration was founded and upheld. We cannot see how one so learned in the law as Mr. Justice Bradley is presumed to be, by reason of his exalted position, can see only the Fourteenth Amendment, as the part of the Constitution, relied on. It is undeniably patent to all that the Thirteenth Amendment more nearly expresses the foundation for the "act." . . .

In October, 1882, five cases were filed or submitted: United States *vs.* Stanley, from Kansas; United States *vs.* Ryan, from California; United States *vs.* Nichols, from Missouri; United States *vs.* Singleton, from New York; and Robinson and wife *vs.* Memphis and Charleston Railroad Company, from Tennessee. Our learned (?) judges occupied a year in considering what their dicta should be. October 15, 1883, found Justice Bradley in his place, on the bench, prepared to voice the opinion of the court as to the rights of more than seven millions of human beings. Mr. Solicitor-General Phillips, had delivered his argument for the life of the law to be maintained. The argument of the Solicitor-General had been supplemented by the eloquent efforts of Mr. William M. Randolph, on behalf of Robinson and wife. Numerous authorities were cited to show that where the Constitution guarantees a right, Congress is empowered to pass the legislation appropriate to give effect to that right. It was also

[2]Sane mind.

maintained and established by judicial precedents, that the constitutionality of the act was not harmed by the nice distinction of "guaranteed
rights," instead of "created rights." Justice Bradley consumes seventeen
pages, to do what in his conscientious (?) opinion he believes to be right.
Justice Harlan, in opposing the position taken by Mr. Bradley, occupies
thirty-seven pages. After reciting the law countenancing the actions insti-
tuted by the sorely aggrieved persons, the first question which propounded itself to the member reading the opinion was, "Are these sections constitutional?" After taking space and time to tell what it is not the
essence of the law to do, the Honorable Judge in *obiter dictum*[3] language
says: "But the responsibility of an independent judgment is now thrown
upon this court; and we are bound to exercise it according to the best
lights we have."

Why this apologetic language? Are we not acquainted with the functions and duties of our court of last resort? Do we not know that the judges
thereof are appointed for life, subject only to their good behavior? This
deciding judge says: "The power is sought, first in the Fourteenth Amendment: and the views and arguments of distinguished senators, advanced
whilst the law was under consideration, claiming authority to pass it by
virtue of that amendment, are the principal arguments adduced in favor
of the power. We have carefully considered those arguments, as was due
to the eminent ability of those who put them forward, and have felt, in all
its force, the weight of authority which always invests a law that Congress
deems itself competent to pass." It is not said that arguments opposed to
the passage of the act were noticed. It is not said that this Honorable
Judge, long before this question of law was brought before him, had pre-
determined its nonconstitutionality.

It is not hinted that this Republican Supreme Court had caused it to
be noised abroad what their "finding" would be if the "law" was inquired
into. The court, it is said, could see, and only see, negroes in Kansas and
Missouri intermingling with white persons in hotels and inns; negroes
in California and New York associating on equal terms with Caucasians
in theaters; and negroes in the presence of those free from the taint of
African blood in the parlor-cars of Tennessee. These sights completely
blinded the eyes of the, at other times, learned judges, and one of their
number, not too full of indignation for utterance, proclaimed aloud,
these things may not be; these pictures shall not in future be produced;
the law is unconstitutional; and all of the other members, save one, said,
amen. Negroes may come as servants into all of the hotels, inns, the-

[3] Relating to an incidental, nonbinding, judicial opinion.

aters and parlor-cars, but they shall never be received as equals — as are other persons. A negro woman with a white baby in her arms may go to the table in the finest and most aristocratic hotel, and there, as a servant, be permitted to associate with all present, of whatever nationality. The same woman, unaccompanied by said baby, or coming without the distinguished rank of servant, is given to understand that she can not enter. And what is more, by the Bradley infamous decision, may be by force of arms prevented from entering. A negro, whose father is a white man, and whose mother's father was white, if marked sufficiently to tell that he is somewhat negro is denied admission into certain places; the same resorts or places of entertainment being readily granted to the inky dark negro who is accompanying an invalid white man. The gambler, cut-throat, thief, despoiler of happy homes and the cowardly assassin need only to have white faces in order to be accommodated with more celerity and respect than are our lawyers, doctors, teachers and humble preachers. Talk about the "Dred Scott" decision; why it was only a mole-hill in comparison with this obstructing Rocky Mountain to the freedom of citizenship. I am charged by your Pennsylvania correspondent with saying that, "By the decision of the Republican Supreme Court colored people may be turned out of hotels, cheated, abused and insulted on steamboats and railroads without legal redress." I am of the opinion that the reporter on your paper who published the above quotation as coming from me, made no mistake unless it was that of making it more mild than I intended. When I use the term, "cheated," I mean that colored persons are required to pay first-class fare and in payment there for are given no-class treatment, or at least the kind which no other human being, paying first-class fare, is served. Some conveyances excepted, I must say to their credit. Bohemians, Scandinavians, Greasers, Italians and Mongolians all precede negroes. When Mr. Justice Harlan shall have retired from the bench by reason of age and infirmity, I pray him to accept, take and carry with him into his retirement the boiled-down essence of the love of more than eight million negroes, who delight to honor an individual whose vertebræ is strong enough to stem the tide of race prejudice. His decision dissenting in favor of equal and exact justice to all men will last always, will never be forgotten as long as there is a descendant of the American negro on the earth: I have no doubt that the feeling of Justice Harlan when seeking rest upon his soft couch on the night of that fateful day in October, was different to the emotions present with Judge Bradley. The latter had doomed seven million human beings and their posterity to "stalls" and "nooks," denoting inferiority: the other had attempted to protect them from American barbarism and vandalism. Seven million persons, many of whom are not only related to Justice Bradley's race by affinity, but by consanguinity, cannot move the bowels of his compassion to the extent of framing or constructing even one sentence in all of that notorious decision which fairly can be interpreted as a friendly regard for the rights of those struggling souls who cried to God, while carrying the burden of bondage for more than two hundred and forty years. God will some day raise up another Lincoln, another Thad. Stevens[4] and another Charles Sumner.[5] In my opinion, if Jesus was on earth, he would say, when speaking of eight members of the Supreme Court and the decision which worked such acerb and cruel wrong upon my people, "Father, forgive them; they know not what they do." . . .

. . . Sane men know that the gentlemen in Congress who voted for this act of 1875 understood full well the condition of our country, as did the powers amending the Constitution abolishing slavery. The intention was to entirely free, not to partly liberate. The desire was to remove the once slave so far from his place of bondage, that he would not even remember it, if such a thing were possible. Congress stepped in and said, he shall vote, he shall serve on juries, he shall testify in court, he shall enter the professions, he shall hold offices, he shall be treated like other men, in all places the conduct of which is regulated by law, he shall in no way be reminded by partial treatment, by discrimination, that he was once a "chattel," a "thing." Certainly Congress had a right to do this. The power that made the slave a man instead of a "thing" had the right to fix his status. The height of absurdity, the chief point in idiocy, the brand of total imbecility, is to say that the Negro shall vote a privilege into existence which one citizen may enjoy for pay, to the exclusion of another, coming in the same way, but clothed in the vesture covering the earth when God first looked upon it. Are colored men to vote grants to railroads upon which they cannot receive equal accommodation? When we ask redress, we are told that the State must first pass a law prohibiting us from enjoying certain privileges and rights, and that after such laws have been passed by the State, we can apply to the United States courts to have such laws declared null and void by *quo warranto* proceedings.[6] The Supreme Court, when applied to, will say to the State, you must not place such laws on your statute book. You can continue your discrimination on account

---

[4]Thaddeus Stevens (1792–1868), congressional representative from Pennsylvania, was an ardent abolitionist and was instrumental in drafting the Fourteenth Amendment.

[5]Charles Sumner (1811–1874) was a Massachusetts senator, lawyer, and abolitionist.

[6]A hearing to determine by what authority a state has power to limit the rights of its citizens.

of color. You can continue to place the badge of slavery on persons having more than one-eighth part of Negro blood in their veins, and so long as your State legislatures do not license you so to do, you are safe. For if they (the Negroes) come to us for redress, we will talk about the autonomy of the State must be held inviolate, referring them back to you for satisfaction.

Do you know of anything more degrading to our country, more damnable? The year after this decision the Republican party met with defeat, because it acquiesced by its silence in that abominable decision, nor did it lift a hand to strike down that diabolical sham of judicial monstrosity, neither in Congress nor the great national convention which nominated Blaine and Logan.[7] God, however, has placed them in power again, using the voters and our manner of electing electors as instruments in his hands. God would have men do right, harm no one, and to render to every man his just due. Mr. Justice Harlan rightly says that the Thirteenth Amendment intended that the white race should have no privilege whatsoever pertaining to citizenship and freedom, that was not alike extended and to be enjoyed by those persons who, though the greater part of them were slaves, were invited by an act of Congress to aid in saving from overthrow a government which, theretofore, by all of its departments, had treated them as an inferior race, with no legal rights or privileges except such as the white race might choose to grant. It is an indisputable fact that the amendment last mentioned may be exerted by legislation of a direct and primary character for the eradication, not simply of the institution of slavery, but of its badges and incidents indicating that the individual was once a slave.

. . . Will the day come when Justice Bradley will want to hide from his decree of the 15th day of October, 1883, and say *non est factum?*[8] I conclude with great reluctance these brief lines, assuring you that the subject is just opened and if desired by you, I will be glad to give it elaborate attention. I ask no rights and privileges for my race in this country, which I would not contend for on behalf of the white people were the conditions changed, or were I to find proscribed white men in Africa where black rules.

A word more and I am done, as you wish brevity. God may forgive this corps of unjust judges, but I never can, their very memories will also be detested by my children's children, nor am I alone in this detestation. The

---

[7]James G. Blaine and John A. Logan were the unsuccessful Republican candidates for president and vice president in 1884.
[8]"It was not done."

eight millions of my race and their posterity will stand horror-frozen at the very mention of their names. The scenes that have passed under my eyes upon the public highways, the brutal treatment of helpless women which I have witnessed, since that decision was proclaimed, is enough to move heaven to tears and raise a loud acclaim in hell over the conquest of wrong. But we will wait and pray, and look for a better day, for God still lives and the LORD OF HOSTS REIGNS.

I am, sir, yours, for the Fatherhood of God, and the Brotherhood of man.

IDA B. WELLS-BARNETT
"The Case Stated"
1895

The late nineteenth and early twentieth centuries witnessed an escalation of white supremacist ideologies and practices. As the pseudoscientific cant of black inferiority gained increasing respectability, so did Jim Crow practices. In this environment, white violence against blacks expanded greatly. The rape of black women by white men during and after slavery, particularly at the turn of the century, drew no comment from whites allegedly concerned with the heinous crime of rape and the sexual virtue of white women. Instead, they focused their attention on what crusading journalist, civil rights activist, and black spokeswoman Ida B. Wells-Barnett denounced as the "threadbare lie" of significant and growing numbers of black men raping white women. In her writings she demonstrated that the racist imagination used this vicious falsehood to justify the lynching of black men by white men and women.

In fact, this slanderous pretext was pivotal to a gruesome system of lynch law that condoned untold numbers of murders of black women, children, and men. Because of their alleged involvement in the rape of white women, the latter constituted the largest category of victims of lynch law. And these crimes against individual blacks were part of a broader system of antiblack violence that included white pogroms against rural and urban blacks. Read

Ida B. Wells-Barnett, "The Case Stated," in *A Red Record*, reprinted in *Southern Horrors and Other Writings: The Anti-Lynching Campaign of Ida B. Wells, 1892–1900*, ed. Jacqueline Jones Royster (Boston: Bedford Books, 1997), 75–82.

against this tragic backdrop, Wells-Barnett's critique is both stinging and telling.

This excerpt is chapter 1 from Wells-Barnett's book A Red Record, *the first social-scientific analysis of the lynch law phenomenon. Wells-Barnett's research grew out of having witnessed the blatant injustice of lynch law in Memphis, Tennessee, where she edited a newspaper, the* Free Speech. *Her heated editorials denouncing lynch law and related injustices led a white mob to destroy her press and forced her to relocate in Chicago. The real issue, she insisted, was the white fear of and anger over black freedom, particularly black success under freedom. Such success constituted an egregious violation of the racist stereotype of black incapacity and its structural corollary: the Negro's "place" within a subordinate caste.*

In this excerpt, she focuses her attention on laying out her argument with respect to the false rape claim against black men. How does she categorize the history of white-on-black violence? What analysis does she make of each period? How does the concept of honor figure in her discussion? How does gender operate in her analysis? What is the connection between lynch law and the black freedom struggle at the turn of the century?

The student of American sociology will find the year 1894 marked by a pronounced awakening of the public conscience to a system of anarchy and outlawry which had grown during a series of ten years to be so common, that scenes of unusual brutality failed to have any visible effect upon the humane sentiments of the people of our land.

Beginning with the emancipation of the Negro, the inevitable result of unbridled power exercised for two and a half centuries, by the white man over the Negro, began to show itself in acts of conscienceless outlawry. During the slave regime, the Southern white man owned the Negro body and soul. It was to his interest to dwarf the soul and preserve the body. Vested with unlimited power over his slave, to subject him to any and all kinds of physical punishment, the white man was still restrained from such punishment as tended to injure the slave by abating his physical powers and thereby reducing his financial worth. While slaves were scourged mercilessly, and in countless cases inhumanly treated in other respects, still the white owner rarely permitted his anger to go so far as to take a life, which would entail upon him a loss of several hundred dollars. The slave was rarely killed, he was too valuable; it was easier and quite as effective, for discipline or revenge, to sell him "Down South."

But Emancipation came and the vested interests of the white man in the Negro's body were lost. The white man had no right to scourge the emancipated Negro, still less has he a right to kill him. But the Southern white people had been educated so long in that school of practice, in which might makes right, that they disdained to draw strict lines of action in dealing with the Negro. In slave times the Negro was kept subservient and submissive by the frequency and severity of the scourging, but, with freedom, a new system of intimidation came into vogue; the Negro was not only whipped and scourged; he was killed.

Not all nor nearly all of the murders done by white men, during the past thirty years in the South, have come to light, but the statistics as gathered and preserved by white men, and which have not been questioned, show that during these years more than ten thousand Negroes have been killed in cold blood, without the formality of judicial trial and legal execution. And yet, as evidence of the absolute impunity with which the white man dares to kill a Negro, the same record shows that during all these years, and for all these murders only three white men have been tried, convicted, and executed. As no white man has been lynched for the murder of colored people, these three executions are the only instances of the death penalty being visited upon white men for murdering Negroes.

Naturally enough the commission of these crimes began to tell upon the public conscience, and the Southern white man, as a tribute to the nineteenth century civilization, was in a manner compelled to give excuses for his barbarism. His excuses have adapted themselves to the emergency, and are aptly outlined by that greatest of all Negroes, Frederick Douglass, in an article of recent date, in which he shows that there have been three distinct eras of Southern barbarism, to account for which three distinct excuses have been made.

The first excuse given to the civilized world for the murder of unoffending Negroes was the necessity of the white man to repress and stamp out alleged "race riots." For years immediately succeeding the war there was an appalling slaughter of colored people, and the wires usually conveyed to northern people and the world the intelligence, first, that an insurrection was being planned by Negroes, which, a few hours later, would prove to have been vigorously resisted by white men, and controlled with a resulting loss of several killed and wounded. It was always a remarkable feature in these insurrections and riots that only Negroes were killed during the rioting, and that all the white men escaped unharmed.

From 1865 to 1872, hundreds of colored men and women were mercilessly murdered and the almost invariable reason assigned was that they met their death by being alleged participants in an insurrection or riot. But this story at last wore itself out. No insurrection ever materialized; no Negro rioter was ever apprehended and proven guilty, and no dynamite ever recorded the black man's protest against oppression and wrong. It was too much to ask thoughtful people to believe this transparent story, and the southern white people at last made up their minds that some other excuse must be had.

Then came the second excuse, which had its birth during the turbulent times of reconstruction. By an amendment to the Constitution the Negro was given the right of franchise, and, theoretically at least, his ballot became his invaluable emblem of citizenship. In a government "of the people, for the people, and by the people," the Negro's vote became an important factor in all matters of state and national politics. But this did not last long. The southern white man would not consider that the Negro had any right which a white man was bound to respect, and the idea of a republican form of government in the southern states grew into general contempt. It was maintained that "This is a white man's government," and regardless of numbers the white man should rule. "No Negro domination" became the new legend on the sanguinary banner of the sunny South, and under it rode the Ku Klux Klan, the Regulators, and the lawless mobs, which for any cause chose to murder one man or a dozen as suited their purpose best. It was a long, gory campaign; the blood chills and the heart almost loses faith in Christianity when one thinks of Yazoo, Hamburg, Edgefield, Copiah, and the countless massacres of defenseless Negroes, whose only crime was the attempt to exercise their right to vote.

But it was a bootless strife for colored people. The government which had made the Negro a citizen found itself unable to protect him. It gave him the right to vote, but denied him the protection which should have maintained that right. Scourged from his home; hunted through the swamps; hung by midnight raiders, and openly murdered in the light of day, the Negro clung to his right of franchise with a heroism which would have wrung admiration from the hearts of savages. He believed that in that small white ballot there was a subtle something which stood for manhood as well as citizenship, and thousands of brave black men went to their graves, exemplifying the one by dying for the other. . . .

Brutality still continued; Negroes were whipped, scourged, exiled, shot and hung whenever and wherever it pleased the white man so to treat them, and as the civilized world with increasing persistency held the white people of the South to account for its outlawry, the murderers invented the third excuse — that Negroes had to be killed to avenge their assaults upon women. There could be framed no possible excuse more harmful to the Negro and more unanswerable if true in its sufficiency for the white man.

Humanity abhors the assailant of womanhood, and this charge upon the Negro at once placed him beyond the pale of human sympathy. With such unanimity, earnestness and apparent candor was this charge made and reiterated that the world has accepted the story that the Negro is a monster which the Southern white man has painted him. And to-day, the Christian world feels, that while lynching is a crime, and lawlessness and anarchy the certain precursors of a nation's fall, it can not by word or deed, extend sympathy or help to a race of outlaws, who might mistake their plea for justice and deem it an excuse for their continued wrongs.

The Negro has suffered much and is willing to suffer more. He recognizes that the wrongs of two centuries can not be righted in a day, and he tries to bear his burden with patience for to-day and be hopeful for to-morrow. But there comes a time when the veriest worm will turn, and the Negro feels to-day that after all the work he has done, all the sacrifices he has made, and all the suffering he has endured, if he did not, now, defend his name and manhood from this vile accusation, he would be unworthy even of the contempt of mankind. It is to this charge he now feels he must make answer.

If the Southern people in defense of their lawlessness, would tell the truth and admit that colored men and women are lynched for almost any offense, from murder to a misdemeanor, there would not now be the necessity for this defense. But when they intentionally, maliciously and constantly belie the record and bolster up these falsehoods by the words of legislators, preachers, governors and bishops, then the Negro must give to the world his side of the awful story.

A word as to the charge itself. In considering the third reason assigned by the Southern white people for the butchery of blacks, the question must be asked, what the white man means when he charges the black man with rape. Does he mean the crime which the statutes of the civilized states describe as such? Not by any means. With the Southern white man, any mesalliance existing between a white woman and a colored man is a sufficient foundation for the charge of rape. The Southern white man says that it is impossible for a voluntary alliance to exist between a white woman and a colored man, and therefore, the fact of an alliance is a proof of force. In numerous instances where colored men have been lynched on the charge of rape, it was positively known at the time of lynching, and indisputably proven after the victim's death, that the relationship sustained between the man and woman was voluntary and clandestine, and that in no court of law could even the charge of assault have been successfully maintained. . . .

During all the years of slavery, no such charge was ever made, not even during the dark days of the rebellion, when the white man, following the fortunes of war went to do battle for the maintenance of slavery. While the master was away fighting to forge the fetters upon the slave, he left his wife and children with no protectors save the Negroes themselves. And yet during those years of trust and peril, no Negro proved recreant to his trust and no white man returned to a home that had been dispoiled.

Likewise during the period of alleged "insurrection," and alarming "race riots," it never occurred to the white man, that his wife and children were in danger of assault. Nor in the Reconstruction era, when the hue and cry was against "Negro Domination," was there ever a thought that the domination would ever contaminate a fireside or strike to death the virtue of womanhood. It must appear strange indeed, to every thoughtful and candid man, that more than a quarter of a century elapsed before the Negro began to show signs of such infamous degeneration.

. . . True chivalry respects all womanhood, and no one who reads the record, as it is written in the faces of the million mulattoes in the South, will for a minute conceive that the southern white man had a very chivalrous regard for the honor due the women of his own race or respect for the womanhood which circumstances placed in his power. That chivalry which is "most sensitive concerning the honor of women" can hope for but little respect from the civilized world, when it confines itself entirely to the women who happen to be white. Virtue knows no color line, and the chivalry which depends upon complexion of skin and texture of hair can command no honest respect.

When emancipation came to the Negroes, there arose in the northern part of the United States an almost divine sentiment among the noblest, purest and best white women of the North, who felt called to a mission to educate and Christianize the millions of southern ex-slaves. From every nook and corner of the North, brave young white women answered that call and left their cultured homes, their happy associations and their lives of ease, and with heroic determination went to the South to carry light and truth to the benighted blacks. . . .They became social outlaws in the South. The peculiar sensitiveness of the southern white men for women, never shed its protecting influence about them. No friendly word from their own race cheered them in their work; no hospitable doors gave them the companionship like that from which they had come. No chivalrous white man doffed his hat in honor or respect. They were "Nigger teachers"—unpardonable offenders in the social ethics of the South, and were insulted, persecuted and ostracised, not by Negroes, but by the white manhood which boasts of its chivalry toward women.

... Before the world adjudges the Negro a moral monster, a vicious assailant of womanhood and a menace to the sacred precincts of home, the colored people ask the consideration of the silent record of gratitude, respect, protection and devotion of the millions of the race in the South, to the thousands of northern white women who have served as teachers and missionaries since the war. ...

It is his regret, that, in his own defense, he must disclose to the world that degree of dehumanizing brutality which fixes upon America the blot of a national crime. Whatever faults and failings other nations may have in their dealings with their own subjects or with other people, no other civilized nation stands condemned before the world with a series of crimes so peculiarly national. It becomes a painful duty of the Negro to reproduce a record which shows that a large portion of the American people avow anarchy, condone murder and defy the contempt of civilization.

These pages are written in no spirit of vindictiveness, for all who give the subject consideration must concede that far too serious is the condition of that civilized government in which the spirit of unrestrained outlawry constantly increases in violence, and casts its blight over a continually growing area of territory. We plead not for the colored people alone, but for all victims of the terrible injustice which puts men and women to death without form of law. During the year 1894, there were 132 persons executed in the United States by due form of law, while in the same year, 197 persons were put to death by mobs who gave the victims no opportunity to make a lawful defense. No comment need be made upon a condition of public sentiment responsible for such alarming results. ...

PAUL LAURENCE DUNBAR
"We Wear the Mask"
1895

Blacks have employed a variety of strategies in their ongoing liberation struggle. On one level, political organizing, protest speeches and marches, boycotts, and legal campaigns represent the vital arena of open confrontation or overt resistance to oppression. This vigorous tradition has been expressed

Paul Laurence Dunbar, "We Wear the Mask," *Complete Works of Paul Laurence Dunbar* (New York: Dodd, Mead, 1922), 71.

*in the antebellum abolitionist movement and the twentieth-century cam-
paign for a federal antilynching law.*

*On another level, there has existed the often far less visible yet deeply
political subterranean world of day-to-day resistance through evasion,
lying, stealth, and ironic action. This masked, typically hidden, often individual mode of resistance is far more difficult to demarcate and to analyze than the comparatively straightforward arena of overt resistance.
Indifferent work, recalcitrance, work slowdowns, and sabotage constitute
undercover forms of subversion. A good example of ironic action was
Booker T. Washington's turn-of-the-century open embrace of Jim Crow at
the same time he was secretly sponsoring court challenges to that very same
caste system.*

*The following poem by Paul Laurence Dunbar (1872–1906) vividly
evokes the undercover world of veiled black resistance. The most celebrated
black literary artist at the turn of the century, Dunbar was best known for
his dialect poetry re-creating the vernacular speech of untutored blacks.
"We Wear the Mask" is radically different and representative of his work in
more traditional poetic forms. That difference is also reflected here in the
poem's thematic content: the subversive possibilities of masking. It speaks
insightfully to the innumerable ways unsung blacks, like oppressed peoples
generally, have manipulated their relative powerlessness in specific situations to their advantage. An example is the ubiquitous practice of stealthily
appropriating the goods and services of well-to-do whites for the benefit of
struggling blacks.*

*This resistance strategy reflects both affirmation and endurance. It has a
political corollary in the infinite number of examples of individual—thus often
unrecorded—acts of black protest against Jim Crow. For example, how are
we to interpret the actions of a poor black woman who appears to acquiesce
to Jim Crow in her daily life but who one day stages a battle for her right to a
seat in the ladies' car of the trolley rather than the segregated black car? Our
heroine's unwritten story, her unprivileged situation—her class, gender, and
race status—as well as scholars' overreliance on traditional historical sources
have combined to render invisible, and thus insignificant, protests like hers.*

*Dunbar's poem demands that we uncover her story, validate her voice,
and expand our understanding of black resistance specifically and the black
freedom struggle generally. Do you see a connection between modes of black
protest—in this instance overt and covert—and the extent of white racism,
especially in the context of turn-of-the-century race relations? What do you
see as the political strengths and weaknesses of both forms of protest? Historically, has covert resistance in any way influenced overt resistance as represented by civil rights struggles?*

We wear the mask that grins and lies,
It hides our cheeks and shades our eyes—
This debt we pay to human guile;
With torn and bleeding hearts we smile,
And mouth with myriad subtleties.

Why should the world be overwise,
In counting all our tears and sighs?
Nay, let them only see us, while
    We wear the mask.

We smile, but, O great Christ, our cries
To Thee from tortured souls arise.
We sing, but oh, the clay is vile
Beneath our feet, and long the mile;
But let the world dream otherwise,
    We wear the mask.

JUSTICE HENRY BROWN

Majority Opinion in Plessy v. Ferguson

1896

In 1892 Homer Plessy, a citizen of Louisiana who was one-eighth black and seven-eighths white, was arrested for riding in a whites-only car during an intrastate trip. The episode had been planned by a committee of eminent African Americans from New Orleans—largely Creoles, or French-speaking people of mixed heritage—as a challenge to a Louisiana statute requiring railway companies to provide separate and equivalent accommodations for whites and blacks. The constitutional issue at stake was the right of a state to make and enforce this kind of racial discrimination. The Supreme Court ruling in the Civil Rights Cases *(1883) had upheld racial discrimination in public accommodations run by private concerns, such as steamship and railroad companies. These rulings invalidated the Civil Rights Act of 1875.*

In the late nineteenth and early twentieth centuries, the state of Louisiana, along with other southern states, passed laws institutionaliz-

Plessy v. Ferguson, 163 U.S. 538 (1896).

ing white supremacy. These state laws extended private forms of racial discrimination to the public sphere, to public action, and to the realm of state mandate and state enforcement. Separate racial arrangements were legally required in sites such as hotels, theaters, courthouses, city halls, parks, stores, and restrooms. These arrangements varied; sometimes they were wholly separate domains, such as separate schools, sometimes segregated sites and facilities within facilities, such as separate and inferior spaces for blacks to eat in restaurants, provided they were served at all.

Plessy's case was part of an important, growing, and increasingly unsuccessful legal challenge among southern blacks and their allies to this pervasive system of state-endorsed racial caste. In 1892, the state supreme court upheld the constitutionality of the Louisiana Jim Crow railway statute; in 1896, the United States Supreme Court endorsed that decision in Plessy v. Ferguson.

Included here are excerpts from the very influential majority opinion issued by Justice Henry Brown. Brown's opinion turned on the argument of the state's power to discriminate against blacks as a function of its constitutionally defined police power. The logic of this argument insisted that state-sanctioned racial segregation codified social custom — that is, the widespread practice of Jim Crow — and promoted the public good, in this case racial harmony. The majority decision also declared that separate and equal provisions for blacks and whites was consistent with the equal protection clause of the Fourteenth Amendment. The decision also represented race as a rational, not arbitrary, classification for the delineation of state-defined rights.

Plessy *was the most widely cited precedent for the legality of Jim Crow until it was overturned by* Brown *in 1954; its authority and impact were extraordinary. In fact, the ruling came to signify the very idea of American apartheid.* Plessy *legitimized the culture of Jim Crow as it spread throughout American life; it simultaneously institutionalized and symbolized the centrality of race in modern America.*

To begin to grapple with the awesome impact of the Plessy *decision, several questions bear consideration. How is race defined and how does it function in the decision? At the time, how did the state legally shape race relations? What were the social, political, and ideological ramifications of separate but equal as theory? as practice? Assess the persuasiveness of the legal argument. What in the history of the first half of the twentieth century in the United States, and indeed the world, helps to explain the fact that* Plessy *lasted for more than fifty years?*

This case turns upon the constitutionality of an act of the General Assembly of the State of Louisiana, passed in 1890, providing for separate railway carriages for the white and colored races. . . .

The constitutionality of this act is attacked upon the ground that it conflicts both with the Thirteenth Amendment of the Constitution, abolishing slavery, and the Fourteenth Amendment, which prohibits certain restrictive legislation on the part of the States.

1. That it does not conflict with the Thirteenth Amendment of the Constitution, which abolished slavery and involuntary servitude, except as a punishment for crime, is too clear for argument. Slavery implies involuntary servitude—a state of bondage; the ownership of mankind as a chattel, or at least the control of the labor and services of one man for the benefit of another, and the absence of a legal right to the disposal of his own person, property and services. This amendment was . . . intended primarily to abolish slavery, as it had been previously known in this country, and . . . equally forbade Mexican peonage or the Chinese coolie trade, when they amounted to slavery or involuntary servitude, and that the use of the word "servitude" was intended to prohibit the use of all forms of involuntary slavery, of whatever class or name. . . .

So, too, in the *Civil Rights cases* . . . it was said that the act of a mere individual, the owner of an inn, a public conveyance or place of amusement, refusing accommodations to colored people, cannot be justly regarded as imposing any badge of slavery or servitude upon the applicant, but only as involving an ordinary civil injury, properly cognizable by the laws of the State, and presumably subject to redress by those laws until the contrary appears. . . .

A statute which implies merely a legal distinction between the white and colored races—a distinction which is founded in the color of the two races, and which must always exist so long as white men are distinguished from the other race by color—has no tendency to destroy the legal equality of the two races, or reestablish a state of involuntary servitude. . . .

2. By the Fourteenth Amendment, all persons born or naturalized in the United States, and subject to the jurisdiction thereof, are made citizens of the United States and of the State wherein they reside; and the States are forbidden from making or enforcing any law which shall abridge the privileges or immunities of citizens of the United States, or shall deprive any person of life, liberty or property without due process of law or deny to any person within their jurisdiction the equal protection of the laws.

The proper construction of this amendment . . . involved, however, not a question of race, but one of exclusive privileges. The case did not call for any expression of opinion as to the exact rights it was intended to secure to the colored race, but it was said generally that its main purpose was to establish the citizenship of the negro; to give definitions of citizenship of the United States and of the States, and to protect from the hostile legislation of the States the privileges and immunities of citizens of the United States, as distinguished from those of citizens of the States.

The object of the amendment was undoubtedly to enforce the absolute equality of the two races before the law, but in the nature of things it could not have been intended to abolish distinctions based upon color, or to enforce social, as distinguished from political equality, or a commingling of the two races upon terms unsatisfactory to either. Laws permitting, and even requiring, their separation in places where they are liable to be brought into contact do not necessarily imply the inferiority of either race to the other, and have been generally, if not universally, recognized as within the competency of the state legislatures in the exercise of their police power. The most common instance of this is connected with the establishment of separate schools for white and colored children, which has been held to be a valid exercise of the legislative power even by courts of States where the political rights of the colored race have been longest and most earnestly enforced. . . .

. . . The power to assign to a particular coach obviously implies the power to determine to which race the passenger belongs, as well as the power to determine who, under the laws of the particular State, is to be deemed a white, and who a colored person. . . .

. . . [I]t is also suggested . . . that the same argument that will justify the state legislature in requiring railways to provide separate accommodations for the two races will also authorize them to require separate cars to be provided for people whose hair is of a certain color, or who are aliens, or who belong to certain nationalities, or to enact laws requiring colored people to walk upon one side of the street, and white people upon the other, or requiring white men's houses to be painted white, and colored men's black, or their vehicles or business signs to be of different colors, upon the theory that one side of the street is as good as the other, or that a house or vehicle of one color is as good as one of another color. The reply to all this is that every exercise of the police power must be reasonable, and extend only to such laws as are enacted in good faith for the promotion for the public good, and not for the annoyance or oppression of a particular class. . . .

So far, then, as a conflict with the Fourteenth Amendment is concerned the case reduces itself to the question whether the statute of Louisiana is a reasonable regulation and with respect to this there must necessarily be a large discretion on the part of the legislature. In determining the question of reasonableness it is at liberty to act with reference to the established usages, customs and traditions of the people, and with a view to the promotion of their comfort, and the preservation of the public peace and good order. Gauged by this standard, we cannot say that a law which authorizes or even requires the separation of the two races in public conveyances is unreasonable, or more obnoxious to the Fourteenth Amendment than the acts of Congress requiring separate schools for colored children in the District of Columbia, the constitutionality of which does not seem to have been questioned, or the corresponding acts of state legislatures.

We consider the underlying fallacy of the plaintiff's argument to consist in the assumption that the enforced separation of the two races stamps the colored race with a badge of inferiority. If this be so, it is not by reason of anything found in the act, but solely because the colored race chooses to put that construction upon it. The argument necessarily assumes that if, as has been more than once the case, and is not unlikely to be so again, the colored race should become the dominant power in the state legislature, and should enact a law in precisely similar terms, it would thereby relegate the white race to an inferior position. We imagine that the white race, at least, would not acquiesce in this assumption. The argument also assumes that social prejudices may be overcome by legislation, and that equal rights cannot be secured to the negro except by an enforced commingling of the two races. We cannot accept this proposition. If the two races are to meet upon terms of social equality, it must be the result of natural affinities, a mutual appreciation of each other's merits and a voluntary consent of individuals. . . . Legislation is powerless to eradicate racial instincts or to abolish distinctions based upon physical differences and the attempt to do so can only result in accentuating the difficulties of the present situation. If the civil and political rights of both races be equal one cannot be inferior to the other civilly or politically. If one race be inferior to the other socially, the Constitution of the United States cannot put them upon the same plane. . . .

JUSTICE JOHN MARSHALL HARLAN

Dissenting Opinion in Plessy v. Ferguson

1896

Harlan saw Brown's majority opinion as a violation of Plessy's constitutional right to equality before the law secured by the Thirteenth and Fourteenth Amendments. In a famous legal moment, he argued for the Constitution as "color-blind"; as he explained, "there is no caste here." This idea has had a powerful impact on debates about the proper way to view the relationship between the ideal and the reality of the Constitution regarding race. Harlan's dissent has also contributed significantly to discussions of the viability of race as a category for delineating legal rights.

Although in crucial ways Harlan reflected the white supremacist views of his times, he nonetheless saw race as an arbitrary and thus impermissible criterion for establishing constitutional rights. He also saw "equal but separate" as an unreasonable exercise of state power: the legal legitimization of state-endorsed forms of invidious racial discrimination. In addition, as he interpreted it, the majority view in Plessy *was more likely to promote racial discord than racial harmony.*

Harlan's dissent proved vital to the ideological struggle against Jim Crow. It became a touchstone in the legal arguments for integration that culminated in Brown. *This dissent thus provided critical legal support for black freedom, equality, and justice. Still, it is necessary to assess the cogency of Harlan's legal arguments. Do they persuade you? What are the salient differences and similarities between his opinion and that of the majority? How does he define race and how does it function in his argument? How do you explain the influence of Harlan's dissent on the modern black civil rights struggle?*

In respect of civil rights, common to all citizens, the Constitution of the United States does not, I think, permit any public authority to know the race of those entitled to be protected in the enjoyment of such rights. Every true man has pride of race, and under appropriate circumstances when the rights of others, his equals before the law, are not to be affected, it is his privilege to express such pride and to take such action based upon

Plessy v. Ferguson, 163 U.S. 538 (1896).

it as to him seems proper. But I deny that any legislative body or judicial tribunal may have regard to the race of citizens when the civil rights of those citizens are involved. Indeed, such legislation, as that here in question, is inconsistent not only with that equality of rights which pertains to citizenship, National and State, but with the personal liberty enjoyed by every one within the United States.

The Thirteenth Amendment does not permit the withholding or the deprivation of any right necessarily inhering in freedom. It not only struck down the institution of slavery as previously existing in the United States, but it prevents the imposition of any burdens or disabilities that constitute badges of slavery or servitude. It decreed universal civil freedom in this country. The court has so adjudged. But that amendment having been found inadequate to the protection of the rights of those who had been in slavery, it was followed by the Fourteenth Amendment, which added greatly to the dignity and glory of American citizenship, and to the security of personal liberty. . . . These two amendments, if enforced according to their true intent and meaning, will protect all the civil rights that pertain to freedom and citizenship. Finally, . . . that no citizen should be denied, on account of his race, the privilege of participating in the political control of his country . . . was declared . . . by the Fifteenth Amendment.

These notable additions to the fundamental law were welcomed by the friends of liberty throughout the world. They removed the race line from our governmental systems. They had, as this court has said, a common purpose, namely, to secure "to a race recently emancipated, a race that through many generations have been held in slavery, all the civil rights that the superior race enjoy." They declared, in legal effect, this court has further said, "that the law in the States shall be the same for the black as for the white; that all persons, whether colored or white, shall stand equal before the laws of the States, and, in regard to the colored race, for whose protection the amendment was primarily designed, that no discrimination shall be made against them by law because of their color." We also said: "The works of the amendment, it is true, are prohibitory, but they contain a necessary implication of a positive immunity, or right, most valuable to the colored race—the right to exemption from unfriendly legislation against them distinctively as colored—exemption from legal discriminations, implying inferiority in civil society, lessening the security of their enjoyment of the rights which others enjoy, and discriminations which are steps towards reducing them to the condition of a subject race." . . .

It was said in argument that the statute of Louisiana does not discriminate against either race, but prescribes a rule applicable alike to white and colored citizens. But this argument does not meet the difficulty.

Every one knows that the statute in question had its origin in the purpose, not so much to exclude white persons from railroad cars occupied by blacks, as to exclude colored people from coaches occupied by or assigned to white persons. Railroad corporations of Louisiana did not make discrimination among whites in the matter of accommodation for travellers. The thing to accomplish was, under the guise of giving equal accommodation for whites and blacks, to compel the latter to keep to themselves while travelling in railroad passenger coaches. No one would be so wanting in candor as to assert the contrary. The fundamental objection, therefore, to the statute is that it interferes with the personal freedom of citizens. "Personal liberty," it has been well said, "consists in the power of locomotion, of changing situation, or removing one's person to whatsoever places one's own inclination may direct, without imprisonment or restraint, unless by due course of law." If a white man and a black man choose to occupy the same public conveyance on a public highway, it is their right to do so, and no government, proceeding alone on grounds of race, can prevent it without infringing the personal liberty of each.

It is one thing for railroad carriers to furnish, or to be required by law to furnish, equal accommodations for all whom they are under a legal duty to carry. It is quite another thing for government to forbid citizens of the white and black races from travelling in the same public conveyance, and to punish officers of railroad companies for permitting persons of the two races to occupy the same passenger coach. If a State can prescribe, as a rule of civil conduct, that whites and blacks shall not travel as passengers in the same railroad coach, why may it not so regulate the use of the streets of its cities and towns as to compel white citizens to keep on one side of a street and black citizens to keep on the other? Why may it not, upon like grounds, punish whites and blacks who ride together in street cars or in open vehicles on a public road or street? Why may it not require sheriffs to assign whites to one side of a court-room and blacks to the other? And why may it not also prohibit the commingling of the two races in the galleries of legislative halls or in public assemblages convened for the consideration of the political questions of the day? Further, if this statute of Louisiana is consistent with the personal liberty of citizens, why may not the State require the separation in railroad coaches of native and naturalized citizens of the United States, or of Protestants and Roman Catholics?

The answer given at the argument to these questions was that regulations of the kind they suggest would be unreasonable, and could not, therefore, stand before the law. Is it meant that the determination of questions of legislative power depends upon the inquiry whether the statute whose validity is questioned is, in the judgment of the courts, a reasonable one, taking all the circumstances into consideration? A statute may be unreasonable merely because a sound public policy forbade its enactment. But I do not understand that the courts have anything to do with the policy or expediency of legislation. A statute may be valid, and yet, upon grounds of public policy, may well be characterized as unreasonable. . . . Statutes must always have a reasonable construction. Sometimes they are to be construed strictly; sometimes, liberally, in order to carry out the legislative will. But however construed, the intent of the legislature is to be respected, if the particular statute in question is valid, although the courts, looking at the public interests, may conceive the statute to be both unreasonable and impolitic. If the power exists to enact a statute, that ends the matter so far as the courts are concerned. The adjudged cases in which statutes have been held to be void, because unreasonable, are those in which the means employed by the legislature were not at all germane to the end to which the legislature was competent.

The white race deems itself to be the dominant race in this country. And so it is, in prestige, in achievements, in education, in wealth and in power. So, I doubt not, it will continue to be for all time, if it remains true to its great heritage and holds fast to the principles of constitutional liberty. But in view of the Constitution, in the eye of the law, there is in this country no superior, dominant, ruling class of citizens. There is no caste here. Our Constitution is color-blind, and neither knows nor tolerates classes among citizens. In respect of civil rights, all citizens are equal before the law. The humblest is the peer of the most powerful. The law regards man as man, and takes no account of his surroundings or of his color when his civil rights as guaranteed by the supreme law of the land are involved. It is, therefore, to be regretted that this high tribunal, the final expositor of the fundamental law of the land, has reached the conclusion that it is competent for a State to regulate the enjoyment by citizens of their civil rights solely upon the basis of race.

In my opinion, the judgment this day rendered will, in time, prove to be quite as pernicious as the decision made by this tribunal in the *Dred Scott* case. It was adjudged in that case that the descendants of Africans who were imported into this country and sold as slaves were not included nor intended to be included under the word "citizens" in the Constitution, and could not claim any of the rights and privileges which that instrument provided for and secured to citizens of the United States; that at the time of the adoption of the Constitution they were "considered as a subordinate and inferior class of beings, who had been subjugated by the dominant race, and, whether emancipated or not, yet remained subject to their authority, and had no rights or privileges but such as those who held the power and the government might choose to grant them." The recent amendments of the Constitution, it was supposed, had eradicated these principles from our institutions. But it seems that we have yet, in some of the States, a dominant race—a superior class of citizens, which assumes to regulate the enjoyment of civil rights, common to all citizens, upon the basis of race. The present decision, it may well be apprehended, will not only stimulate aggressions, more or less brutal and irritating, upon the admitted rights of colored citizens, but will encourage the belief that it is possible, by means of state enactments, to defeat the beneficent purposes which the people of the United States had in view when they adopted the recent amendments of the Constitution, by one of which the blacks of this country were made citizens of the United States and of the States in which they respectively reside, and whose privileges and immunities, as citizens, the States are forbidden to abridge. Sixty millions of whites are in no danger from the presence here of eight millions of blacks. The destinies of the two races, in this country, are indissolubly linked together, and the interests of both require that the common government of all shall not permit the seeds of race hate to be planted under the sanction of law. What can more certainly arouse race hate, what more certainly create and perpetuate a feeling of distrust between these races, than state enactments, which, in fact, proceed on the ground that colored citizens are so inferior and degraded that they cannot be allowed to sit in public coaches occupied by white citizens? That, as all will admit, is the real meaning of such legislation as was enacted in Louisiana.

The sure guarantee of the peace and security of each race is the clear, distinct, unconditional recognition by our governments, National and State, of every right that inheres in civil freedom, and of the equality before the law of all citizens of the United States without regard to race. State enactments regulating the enjoyment of civil rights, upon the basis of race, and cunningly devised to defeat legitimated results of the war, under the pretence of recognizing equality of rights, can have no other result than to render permanent peace impossible, and to keep alive a conflict of races, the continuance of which must do harm to all concerned. This question is not met by the suggestion that social equality cannot exist between the white and black races in this country. That argument, if it can be properly regarded as one, is scarcely worthy of consideration; for social equality no more exists between two races when travelling in a passenger coach or a public highway than when members of the same races sit by each other in a street car or in the jury box, or stand or sit with each other in a political assembly, or when they use in common the streets of a city of town, or when they are in the same room for the purpose of having their names placed on the registry of voters, or when they approach the ballot-box in order to exercise the high privilege of voting.

There is a race so different from our own that we do not permit those belonging to it to become citizens of the United States. . . . I allude to the Chinese race. But by the statute in question, a Chinaman can ride in the same passenger coach with white citizens of the United States, while citizens of the black race in Louisiana, many of whom, perhaps, risked their lives for the preservation of the Union, who are entitled, by law, to participate in the political control of the State and nation, who are not excluded, by law or by reason of their race, from public stations of any kind, and who have all the legal rights that belong to white citizens, are yet declared to be criminals, liable to imprisonment, if they ride in a public coach occupied by citizens of the white race. . . .

The arbitrary separation of citizens, on the basis of race, while they are on a public highway, is a badge of servitude wholly inconsistent with the civil freedom and the equality before the law established by the Constitution. It cannot be justified upon any legal grounds. . . .

3

Sweatt v. Painter (1950) and *McLaurin v. Oklahoma State Regents* (1950)

"Letters of Negro Migrants of 1916–1918"

1919

During World War I, more than 500,000 southern blacks left for the North. Even more left for the West as well as the North during World War II. These extraordinary mass migrations — the first and second Great Migrations — significantly expanded the black populations in the North and the West. As a result, the black freedom struggle and the issue of black-white relations increasingly became a national rather than a southern regional concern. Indeed, these large-scale migrations profoundly altered twentieth-century United States life and culture.

The following letters are a small slice of a voluminous outpouring from southern blacks writing to the Chicago Defender, *the most widely circulated black newspaper at the time, which vigorously touted the benefits of black migration to the North. The* Defender's *spirited promotional campaign signified a modern reworking of the North Star legend of the slavery era: the North as the land of freedom and opportunity for blacks. Pushed to leave an increasingly racist and repressive South whose cotton-dominated economy was being devastated by the boll weevil blight, black migrants were pulled northward by the lure of wartime jobs and the dream of a better life generally, including better educational opportunities. Unfortunately, as most would soon learn, by various measures blacks might be relatively better off in the North, but it was certainly no haven. There, too, they faced institutionalized as well as interpersonal patterns of prejudice and discrimination.*

Emmet J. Scott, ed., "Letters of Negro Migrants of 1916–1918," *Journal of Negro History* 4 (1919): 304, 317, 320, 413, 419–20, 437.

These letters speak to the migrants' most basic concerns. Why do you think the vast majority of blacks remained in the South? Do you see the first Great Migration as a form of resistance? What do you see as the major ramifications for the black civil rights struggle of the first Great Migration?

LEXINGTON, MISS., May 12–17.

My dear Mr. H——: — I am writing to you for some information and assistance if you can give it.

I am a young man and am disable, in a very great degree, to do hard manual labor. I was educated at Alcorn College and have been teaching a few years : but ah : me the Superintendent under whom we poor colored teachers have to teach cares less for a colored man than he does for the vilest beast. I am compelled to teach 150 children without any assistance and receives only $27.00 a month, the white with 30 get $100.

I am so sick I am so tired of such conditions that I sometime think that life for me is not worth while and most eminently believe with Patrick Henry "Give me liberty or give me death." If I was a strong able bodied man I would have gone from here long ago, but this handicaps me and, I must make inquiries before I leap.

Mr. H——, do you think you can assist me to a position I am good at stenography typewriting and bookkeeping or any kind of work not to rough or heavy. I am 4 feet 6 in high and weigh 105 pounds.

I will gladly give any other information you may desire and will greatly appreciate any assistance you may render me.

SELMA, ALA., May 19, 1917.

Dear Sir: I am a reader of the Chicago Defender I think it is one of the Most Wonderful Papers of our race printed. Sirs I am writeing to see if You all will please get me a job. And Sir I can wash dishes, wash iron nursing work in groceries and dry good stores. Just any of these I can do. Sir, who so ever you get the job from please tell them to send me a ticket and I will pay them. When I get their as I have not got enough money to pay my way. I am a girl of 17 years old and in the 8 grade at Knox Academy School. But on account of not having money enough I had to stop school. Sir I will thank you all with all my heart. May God Bless you all. Please answer in return mail.

MOBILE, ALA., May 11, 1917.

Dear sir and brother: on last Sunday I addressed you a letter asking you for information and I have received no answer. but we would like to know could 300 or 500 men and women get employment? and will the company or thoes that needs help send them a ticket or a pass and let them pay it back in weekly payments? We have men and women here in all lines of work we have organized a association to help them through you.

We are anxiously awaiting your reply.

ALEXANDRIA, LA., June 6, 1917.

Dear Sirs: I am writeing to you all asking a favor of you all. I am a girl of seventeen. School has just closed I have been going to school for nine months and I now feel like I aught to go to work. And I would like very very well for you all to please forward me to a good job. but there isnt a thing here for me to do, the wages here is from a dollar and a half a week. What could I earn Nothing. I have a mother and father my father do all he can for me but it is so hard. A child with any respect about her self or his self wouldnt like to see there mother and father work so hard and earn nothing I feel it my duty to help. I would like for you all to get me a good job and as I havent any money to come on please send me a pass and I would work and pay every cent of it back and get me a good quite place to stay. My father have been getting the defender for three or four months but for the last two weeks we have failed to get it. I dont know why. I am tired of down hear in this———/ I am afraid to say. Father seem to care and then again dont seem to but Mother and I am tired tired of all of this I wrote to you all because I believe you will help I need your help hopeing to here from you all very soon.

AUGUSTA, GA., April 27, 1917.

Sir: Being a constant reader of your paper, I thought of no one better than you to write for information.

I'm desirous of leaving the south but before so doing I want to be sure of a job before pulling out. I'm a member of the race, a normal and colloege school graduate, a man of a family and can give reference. Confidentially this communication between you and me is to be kept a secret.

My children I wished to be educated in a different community than here. Where the school facilities are better and less prejudice shown and in fact where advantages are better for our people in all respect. At present I have a good position but I desire to leave the south. A good position even tho' its a laborer's job paying $4.50 or $5.00 a day will suit me till I can do better. Let it be a job there or any where else in the country, just is it is east or west. I'm quite sure you can put me in touch with some one. I'm a letter carrier now and am also a druggist by profession. Perhaps I may through your influence get a transfer to some eastern or western city.

Nevada or California as western states, I prefer, and I must say that I have nothing against Detroit, Mich.

I shall expect an early reply. Remember keep this a secret please until I can perfect some arrangements.

LANGSTON HUGHES
"I, Too"
1926

The enduring hope underlying the black freedom struggle, obstacles notwithstanding, is captured beautifully in the following poem by Langston Hughes (1902–1967). A popular and prolific man of letters of uncommon range and piercing insight, Hughes possessed a clear-sighted and powerful vision of the centrality of African Americans to the American experience. The same undaunted spirit of protest and struggle animating the poem's protagonist invigorated the NAACP's legal campaign against Jim Crow. Likewise, the moral indignation and integrationism of the poem reflect parallel features of that campaign and the NAACP itself.

White Americans, the poem maintains, would be forced to accept black Americans once the former understood how fundamentally alike they are — how deeply American they both are. Do you agree with this argument? Explain your position. Are you persuaded by the argument of African American identity as essentially American? How did this "Americanism" inform legalism as well as other strategies within the black freedom struggle? How does resistance here compare with resistance as discussed in Paul Laurence Dunbar's poem "We Wear the Mask" (p. 74)?

I, too, sing America.

I am the darker brother.
They send me to eat in the kitchen
When company comes,

Langston Hughes, "I, Too," *Selected Poems* (New York: Vintage, 1990), 275.

But I laugh,
And eat well,
And grow strong.

Tomorrow,
I'll be at the table
When company comes.
Nobody'll dare
Say to me,
"Eat in the kitchen,"
Then.

Besides,
They'll see how beautiful I am
And be ashamed—

I, too, am America.

W. E. B. DU BOIS

"Does the Negro Need Separate Schools?"
1935

The discrimination confronting black schools in the South and antiblack dis-crimination in mixed schools in the North engendered extensive debate among blacks and their allies about how to improve black educational opportunity. Should blacks fight for truly equal, albeit separate, schools— that is, should they seek to make Jim Crow schools equal in fact? Or should they focus on getting rid of separate schools and creating desegregated schools? In the 1930s, W. E. B. Du Bois, the gifted scholar and activist, argued for the former. In the following article on the perils and prospects of separate schools, he argues forcefully that these institutions need to be strengthened intellectually and fiscally and given full black support.

Fully aware of the intraracial as well as racist constraints circumscribing black education at the time, he argues for a policy known as equalization,

W. E. B. Du Bois, "Does the Negro Need Separate Schools?" *Journal of Negro Education*, July 1935, 328–35.

a strategy that he sees as a means toward full educational opportunity for blacks, not as a goal in and of itself. For Du Bois, equalization is the most viable of the available choices, speaking forcefully to the need for black pride in and support of black institutions like schools. He emphasizes that the ultimate goal has to be an integrated and multicultural educational experience truly reflective of the diversity of the American experience. Until race relations rendered this goal attainable, first-rate all-black schools would suffice. Paradoxically, then, he urges the promotion of both blackness and egalitarianism. What are the principal aspects of his overall argument? Do you find his major proposition and supporting arguments persuasive? Explain. What do you see as the ramifications of this principled yet pragmatic position for mid-1930s black civil rights activism? How does this position jibe with that of the NAACP?

There are in the United States some four million Negroes of school age, of whom two million are in school, and of these, four-fifths are taught by forty-eight thousand Negro teachers in separate schools. Less than a half million are in mixed schools in the North, where they are taught almost exclusively by white teachers. Beside this, there are seventy-nine Negro universities and colleges with one thousand colored teachers, beside a number of private secondary schools.

The question which I am discussing is: Are these separate schools and institutions needed? And the answer, to my mind, is perfectly clear. They are needed just so far as they are necessary for the proper education of the Negro race. The proper education of any people includes sympathetic touch between teacher and pupil; knowledge on the part of the teacher, not simply of the individual taught, but of his surroundings and background, and the history of his class and group; such contact between pupils, and between teacher and pupil, on the basis of perfect social equality, as will increase this sympathy and knowledge; facilities for education in equipment and housing, and the promotion of such extra-curricular activities as will tend to induct the child into life.

If this is true, and if we recognize the present attitude of white America toward black America, then the Negro not only needs the vast majority of these schools, but it is a grave question if, in the near future, he will not need more such schools, both to take care of his natural increase, and to defend him against the growing animosity of the whites. It is of course fashionable and popular to deny this; to try to deceive ourselves into thinking that race prejudice in the United States across the Color Line is gradually softening and that slowly but surely we are coming to the time when racial animosities and class lines will be so obliterated that separate schools will be anachronisms.

Certainly, I shall welcome such a time. Just as long as Negroes are taught in Negro schools and whites in white schools; the poor in the slums, and the rich in private schools; just as long as it is impracticable to welcome Negro students to Harvard, Yale and Princeton; just as long as colleges like Williams, Amherst and Wellesley tend to become the property of certain wealthy families, where Jews are not solicited; just so long we shall lack in America that sort of public education which will create the intelligent basis of a real democracy.

Much as I would like this, and hard as I have striven and shall strive to help realize it, I am no fool; and I know that race prejudice in the United States today is such that most Negroes cannot receive proper education in white institutions. If the public schools of Atlanta, Nashville, New Orleans and Jacksonville were thrown open to all races tomorrow, the education that colored children would get in them would be worse than pitiable. It would not be education. And in the same way, there are many public school systems in the North where Negroes are admitted and tolerated, but they are not educated; they are crucified. There are certain Northern universities where Negro students, no matter what their ability, desert, or accomplishment, cannot get fair recognition, either in classroom or on the campus, in dining halls and student activities, or in common human courtesy. It is well-known that in certain faculties of the University of Chicago, no Negro has yet received the doctorate and seldom can achieve the mastership in arts; at Harvard, Yale and Columbia, Negroes are admitted but not welcomed; while in other institutions, like Princeton, they cannot even enroll.

Under such circumstances, there is no room for argument as to whether the Negro needs separate schools or not. The plain fact faces us, that either he will have separate schools or he will not be educated. . . .

There are undoubtedly cases where a minority of leaders force their opinions upon a majority, and induce a community to establish separate schools, when as a matter of fact, there is no general demand for it; there has been no friction in the schools; and Negro children have been decently treated. In that case, a firm and intelligent appeal to public opinion would eventually settle the matter. But the futile attempt to compel even by law a group to do what it is determined not to do, is a silly waste of money, time, and temper.

On the other hand, there are also cases where there has been no separation in schools and no movement toward it. And yet the treatment of Negro children in the schools, the kind of teaching and the kind of advice

they get, is such that they ought to demand either a thorough-going revolution in the official attitude toward Negro students, or absolute separation in educational facilities. To endure bad schools and wrong education because the schools are "mixed" is a costly if not fatal mistake. I have long been convinced, for instance, that the Negroes in the public schools of Harlem are not getting an education that is in any sense comparable in efficiency, discipline, and human development with that which Negroes are getting in the separate public schools of Washington, D.C. And yet on its school situation, black Harlem is dumb and complacent, if not actually laudatory.

Recognizing the fact that for the vast majority of colored students in elementary, secondary, and collegiate education, there must be today separate educational institutions because of an attitude on the part of the white people which is not going materially to change in our time, our customary attitude toward these separate schools must be absolutely and definitely changed. As it is today, American Negroes almost universally disparage their own schools. They look down upon them; they often treat the Negro teachers in them with contempt; they refuse to work for their adequate support; and they refuse to join public movements to increase their efficiency.

These images—a blacks-only back entrance to the all-black balcony in a 1930s Pensacola, Florida, movie theater (whites entered in the front and sat downstairs) and the ubiquitous separate drinking fountains *(facing page)*—illustrate a generic Jim Crow pattern. The stringent rule of racial segregation had to be enforced, even in social sites jointly occupied by blacks and whites. To minimize interracial contact, these locations were at once racialized and spatially partitioned: racially separate (and typically unequal) areas and facilities within the same site.

The stark front and overall impression of poverty represented by this rural all-black schoolhouse contrasts vividly with the bucolic physical setting. Most important, the vibrant world of black schoolchildren at play provides a poignant counterpoint to Jim Crow's harshness and mean-spiritedness.

The reason for this is quite clear, and may be divided into two parts: (1) the fear that any movement which implies segregation even as a temporary, much less as a relatively permanent institution, in the United States, is a fatal surrender of principle, which in the end will rebound and bring more evils on the Negro than he suffers today. (2) The other reason is at bottom an utter lack of faith on the part of Negroes that their race can do anything really well. If Negroes could conceive that Negroes could establish schools quite as good as or even superior to white schools; if Negro colleges were of equal grade in accomplishment and in scientific work with white colleges; then separation would be a passing incident and not a permanent evil; but as long as American Negroes believe that their race is constitutionally and permanently inferior to white people, they necessarily disbelieve in every possible Negro Institution. . . .

. . . I have repeatedly seen wise and loving colored parents take infinite pains to force their little children into schools where the white children, white teachers, and white parents despised and resented the dark child, made mock of it, neglected or bullied it, and literally rendered its life a living hell. Such parents want their child to "fight" this thing out, — but, dear God, at what a cost! Sometimes, to be sure, the child triumphs and teaches the school community a lesson; but even in such cases, the cost may be high, and the child's whole life turned into an effort to win cheap applause at the expense of healthy individuality. . . . Therefore, in evaluating the advantage and disadvantage of accepting race hatred as a brutal but real fact, or of using a little child as a battering ram upon which its nastiness can be thrust, we must give greater value and greater emphasis to the rights of the child's own soul. We shall get a finer, better balance of spirit; an infinitely more capable and rounded personality by putting children in schools where they are wanted, and where they are happy and inspired, than in thrusting them into hells where they are ridiculed and hated. . . .

As long as the Negro student wishes to graduate from Columbia, not because Columbia is an institution of learning, but because it is attended by white students; as long as a Negro student is ashamed to attend Fisk or Howard because these institutions are largely run by black folk, just so long the main problem of Negro education will not be segregation but self-knowledge and self-respect. . . .

. . . This state of mind is suicidal and must be fought, and fought doggedly and bitterly: first, by giving Negro teachers decent wages, decent schoolhouses and equipment, and reasonable chances for advancement; and then by kicking out and leaving to the mercy of the white world those who do not and cannot believe in their own. . . .

The N.A.A.C.P. and other Negro organizations have spent thousands of dollars to prevent the establishment of segregated Negro schools, but scarcely a single cent to see that the division of funds between white and Negro schools, North and South, is carried out with some faint approximation of justice. There can be no doubt that if the Supreme Court were overwhelmed with cases where the blatant and impudent discrimination against Negro education is openly acknowledged, it would be compelled to hand down decisions which would make this discrimination impossible. We Negroes do not dare to press this point and force these decisions because, forsooth, it would acknowledge the fact of separate schools, a fact that does not need to be acknowledged, and will not need to be for two centuries.

Howard, Fisk, and Atlanta are naturally unable to do the type and grade of graduate work which is done at Columbia, Chicago, and Harvard; but why attribute this to a defect in the Negro race, and not to the fact that the large white colleges have from one hundred to one thousand times the funds for equipment and research that Negro colleges can command? To this, it may logically be answered, all the more reason that Negroes should try to get into better-equipped schools, and who pray denies this? But the opportunity for such entrance is becoming more and more difficult, and the training offered less and less suited to the American Negro of today. Conceive a Negro teaching in a Southern school the economics which he learned at the Harvard Business School! Conceive a Negro teacher of history retailing to his black students the sort of history that is taught at the University of Chicago! Imagine the history of Reconstruction being handed by a colored professor from the lips of Columbia professors to the ears of the black belt![1] The results of this kind of thing are often fantastic, and call for Negro history and sociology, and even physical science taught by men who understand their audience, and are not afraid of the truth.

There was a time when the ability of Negro brains to do first-class work had to be proven by facts and figures, and I was a part of the movement that sought to set the accomplishments of Negro ability before the world. But the world before which I was setting this proof was a disbelieving white world. I did not need the proof for myself. I did not dream that my fellow Negroes needed it; but in the last few years, I have become curiously convinced that until American Negroes believe in their own power and ability, they are going to be helpless before the white world, and the white world, realizing this inner paralysis and lack of self-confidence, is going to persist in its insane determination to rule the universe for its own selfish advantage.

Does the Negro need separate schools? God knows he does. But what he needs more than separate schools is a firm and unshakable belief that twelve million American Negroes have the inborn capacity to accomplish just as much as any nation of twelve million anywhere in the world ever accomplished, and that this is not because they are Negroes but because they are human.

So far, I have noted chiefly negative arguments for separate Negro institutions of learning based on the fact that in the majority of cases Negroes are not welcomed in public schools and universities nor treated as fellow human beings. But beyond this, there are certain positive reasons due to the fact that American Negroes have, because of their history, group experiences and memories, a distinct entity, whose spirit

---

[1] This refers to the Cotton South, where a majority of blacks toiled and lived.

and reactions demand a certain type of education for its development. . . .

. . . Negroes must know the history of the Negro race in America, and this they will seldom get in white institutions. Their children ought to study textbooks like Brawley's "Short History," the first edition of Woodson's "Negro in Our History," and Cromwell, Turner and Dykes' "Readings from Negro Authors." Negroes who celebrate the birthdays of Washington and Lincoln, and the worthy, but colorless and relatively unimportant "founders" of various Negro colleges, ought not to forget the 5th of March,—that first national holiday of this country, which commemorates the martyrdom of Crispus Attucks.[2] They ought to celebrate Negro Health Week and Negro History Week. They ought to study intelligently and from their own point of view, the slave trade, slavery, emancipation, Reconstruction, and present economic development.

Beyond this, Negro colleges ought to be studying anthropology, psychology, and the social sciences, from the point of view of the colored races. Today, the anthropology that is being taught, and the expeditions financed for archeological and ethnographical explorations, are for the most part straining every nerve to erase the history of black folk from the record. . . .

. . . But much more is necessary and demanded of Negro scholarship. In history and the social sciences the Negro school and college has an unusual opportunity and role. It does not consist simply in trying to parallel the history of white folk with similar boasting about black and brown folk, but rather an honest evaluation of human effort and accomplishment, without color blindness, and without transforming history into a record of dynasties and prodigies. . . .

Thus, instead of our schools being simply separate schools, forced on us by grim necessity, they can become centers of a new and beautiful effort at human education, which may easily lead and guide the world in many important and valuable aspects. It is for this reason that when our schools are separate, the control of the teaching force, the expenditure of money, the choice of textbooks, the discipline and other administrative matters of this sort ought, also, to come into our hands, and be incessantly demanded and guarded. . . .

I know that this article will forthwith be interpreted by certain illiterate "nitwits" as a plea for segregated Negro schools and colleges. It is not.

---

[2]A black former slave, Crispus Attucks, was one of five men killed by British troops on March 5, 1770, in what became known as the Boston Massacre.

It is simply calling a spade a spade. It is saying in plain English: that a separate Negro school, where children are treated like human beings, trained by teachers of their own race, who know what it means to be black in the year of salvation 1935, is infinitely better than making our boys and girls doormats to be spit and trampled upon and lied to by ignorant social climbers, whose sole claim to superiority is ability to kick "niggers" when they are down. I say, too, that certain studies and discipline necessary to Negroes can seldom be found in white schools.

It means this, and nothing more. . . .

Ten year-old Linda Carol Brown — the "named plaintiff" in *Brown v. Board of Education* — is captured with her six-year-old sister Terry Lynn on their way to the all-black Monroe School in Topeka, Kansas, a little over a year before the momentous 1954 Supreme Court decision in their favor. Unable to attend the nearby, all-white Sumner School because of Jim Crow restrictions, Linda and Terry were forced to navigate the hazardous Rock Island Railroad Switchyard to get to the bus stop for the ride to Monroe. It is worth noting that in 1997, Linda Brown Buckner is engaged in litigation seeking once again to integrate Topeka's public schools.

GUNNAR MYRDAL

From An American Dilemma:
The Negro Problem and Modern Democracy
1944

During World War II, Swedish economist Gunnar Myrdal (1898–1987) led a team of scholars in a social-scientific investigation of America's race problem. The results, published as An American Dilemma, *confirmed an emerging national consensus regarding the centrality of America's race problem to its past, present, and future. Intellectually, the work embraced egalitarianism and rejected racism. Ideologically, it provided a critical historical understanding of the discrepancy between the American creed and America's treatment of its black population. Politically, it stressed the incompatibility between Jim Crow and white supremacy on one hand and the United States' role as a principal global power on the other, especially at a time when peoples of color throughout the world were fighting for independence from European domination. This incomparable work represented and propelled the United States' growing determination to alleviate its race relations quagmire.*

Being both a first-rate social scientist and an outsider to the American scene, Myrdal personified scholarly detachment and objectivity. In most eyes, Myrdal's outsider status enhanced the work's substance, insight, balance, and believability. Do you agree with this assessment? What were the study's major findings? Does anything here surprise you? Overall, is the discussion persuasive? What do you see as the links between this study and the wartime and postwar black civil rights struggle?

. . . *The American Negro problem is a problem in the heart of the American. It is there that the interracial tension has its focus. It is there that the decisive struggle goes on. This is the central viewpoint of this treatise. Though our study includes economic, social, and political race relations, at bottom our problem is the moral dilemma of the American — the conflict between his moral valuations on various levels of consciousness and generality. The "American Dilemma," referred to in the title of this book, is the*

Gunnar Myrdal, *An American Dilemma: The Negro Problem and Modern Democracy* (New York: Harper, 1944), xlvi–lv, 1021–23.

ever-raging conflict between, on the one hand, the valuations preserved on the general plane which we shall call the "American Creed," where the American thinks, talks, and acts under the influence of high national and Christian precepts, and, on the other hand, the valuations on specific planes of individual and group living, where personal and local interests; economic, social, and sexual jealousies; considerations of community prestige and conformity; group prejudice against particular persons or types of people; and all sorts of miscellaneous wants, impulses, and habits dominate his outlook. . . .

Valuations and Beliefs

The Negro problem in America would be of a different nature, and, indeed, would be simpler to handle scientifically, if the moral conflict raged only between valuations held by different persons and groups of persons. The essence of the moral situation is, however, that the conflicting valuations are also held by the same person. *The moral struggle goes on within people and not only between them. As people's valuations are conflicting, behavior normally becomes a moral compromise. There are no homogeneous "attitudes" behind human behavior but a mesh of struggling inclinations, interests, and ideals, some held conscious and some suppressed for long intervals but all active in bending behavior in their direction.*

The unity of a culture consists in the fact that all valuations are mutually shared in some degree. We shall find that even a poor and uneducated white person in some isolated and backward rural region in the Deep South, who is violently prejudiced against the Negro and intent upon depriving him of civic rights and human independence, has also a whole compartment in his valuation sphere housing the entire American Creed of liberty, equality, justice, and fair opportunity for everybody. He is actually also a good Christian and honestly devoted to the ideals of human brotherhood and the Golden Rule. And these more general valuations — more general in the sense that they refer to all human beings — are, to some extent, effective in shaping his behavior. Indeed, it would be impossible to understand why the Negro does not fare worse in some regions of America if it were not constantly kept in mind that behavior is the outcome of a compromise between valuations, among which the equalitarian ideal is one. At the other end, there are few liberals, even in New England, who have not a well-furnished compartment of race prejudice, even if it is usually suppressed from conscious attention. Even the American Negroes share in this community of valuations: they have eagerly imbibed the American Creed and the revolutionary

Christian teaching of common brotherhood; under closer study, they usually reveal also that they hold something of the majority prejudice against their own kind and its characteristics.

The intensities and proportions in which these conflicting valuations are present vary considerably from one American to another, and within the same individual, from one situation to another. The cultural unity of the nation consists, however, in the fact that *most Americans have most valuations in common* though they are arranged differently in the sphere of valuations of different individuals and groups and bear different intensity coefficients. This cultural unity is the indispensable basis for discussion between persons and groups. It is the floor upon which the democratic process goes on.

In America as everywhere else people agree, as an abstract proposition, that *the more general valuations—those which refer to man as such and not to any particular group or temporary situation—are morally higher.* These valuations are also given the sanction of religion and national legislation. They are incorporated into the American Creed. The other valuations—which refer to various smaller groups of mankind or to particular occasions—are commonly referred to as "irrational" or "prejudiced," sometimes even by people who express and stress them. They are defended in terms of tradition, expediency or utility.

Trying to defend their behavior to others, and primarily to themselves, people will attempt to conceal the conflict between their different valuations of what is desirable and undesirable, right or wrong, by keeping away some valuations from awareness and by focusing attention on others. For the same opportune purpose, *people will twist and mutilate their beliefs of how social reality actually is.* In our study we encounter whole systems of firmly entrenched popular beliefs concerning the Negro and his relations to the larger society, which are bluntly false and which can only be understood when we remember the opportunistic *ad hoc* purposes they serve. These "popular theories," because of the rationalizing function they serve, are heavily loaded with emotions. But people also want to be rational. Scientific truth-seeking and education are slowly rectifying the beliefs and thereby also influencing the valuations. In a rationalistic civilization it is not only that the beliefs are shaped by the valuations, but also that the valuations depend upon the beliefs. . . .

It is sometimes assumed to be the mark of "sound" research to disregard the fact that people are moral beings and that they are struggling for their conscience. In our view, this is a bias and a blindness, dangerous to the possibility of enabling scientific study to arrive at true knowledge. Every social study must have its center in an investigation of people's conflicting valuations and their opportune beliefs. They are social facts and can be observed by direct and indirect manifestations. We are, of course, also interested in discovering how these inclinations and loyalties came about and what the factors are upon which they rest. We want to keep free, however, at least at the outset, from any preconceived doctrine or theory, whether of the type making biological characteristics, or economic interests, sexual complexes, power relations, or anything else, the "ultimate" or "basic" cause of these valuations. We hope to come out with a type of systematic understanding as eclectic as common sense itself when it is open-minded.

When we thus choose to view the Negro problem as primarily a moral issue, we are in line with popular thinking. It is as a moral issue that this problem presents itself in the daily life of ordinary people; it is as a moral issue that they brood over it in their thoughtful moments. It is in terms of conflicting moral valuations that it is discussed in church and school, in the family circle, in the workshop, on the street corner, as well as in the press, over the radio, in trade union meetings, in the state legislatures, the Congress and the Supreme Court. The social scientist, in his effort to lay bare concealed truths and to become maximally useful in guiding practical and political action, is prudent when, in the approach to a problem, he sticks as closely as possible to the common man's ideas and formulations, even though he knows that further investigation will carry him into tracts uncharted in the popular consciousness. There is a pragmatic common sense in people's ideas about themselves and their worries, which we cannot afford to miss when we start out to explore social reality. Otherwise we are often too easily distracted by our learned arbitrariness and our pet theories, concepts, and hypotheses, not to mention our barbarous terminology, which we generally are tempted to mistake for something more than mere words. *Throughout this study we will constantly take our starting point in the ordinary man's own ideas, doctrines, theories and mental constructs. . . .*

A White Man's Problem

Although the Negro problem is a moral issue both to Negroes and to whites in America, we shall in this book have to give *primary* attention to what goes on in the minds of white Americans. To explain this direction of our interest a general conclusion from our studies needs to be stated at this point. When the present investigator started his inquiry, his preconception was that it had to be focused on the Negro people and their peculiarities. This is understandable since, from a superficial view, Negro Americans, not only in physical appearance, but also in thoughts, feelings, and in manner of life, seemed stranger to him than did white Americans. Furthermore, most of the literature on the Negro problem dealt with the Negroes: their racial and cultural characteristics, their living standards and occupational pursuits, their stratification in social classes, their migration, their family organization, their religion, their illiteracy, delinquency and disease, and so on. But as he proceeded in his studies into the Negro problem, it became increasingly evident that little, if anything, could be scientifically explained in terms of the peculiarities of the Negroes themselves.

As a matter of fact, in their basic human traits the Negroes are inherently not much different from other people. Neither are, incidentally, the white Americans. But Negroes and whites in the United States live in singular human relations with each other. All the circumstances of life—the "environmental" conditions in the broadest meaning of that term—diverge more from the "normal" for the Negroes than for the whites, if only because of the statistical fact that the Negroes are the smaller group. The average Negro must experience many times more of the "abnormal" interracial relations than the average white man in America.[1] The more important fact, however, is that practically all the economic, social, and political power is held by whites. The Negroes do not by far have anything approaching a tenth of the things worth having in America.

It is thus the white majority group that naturally determines the Negro's "place." All our attempts to reach scientific explanations of why the Negroes are what they are and why they live as they do have regularly led to determinants on the white side of the race line. In the practical and political struggles of effecting changes, the views and attitudes of the white Americans are likewise strategic. The Negro's entire life, and, consequently, also his opinions on the Negro problem, are, in the main, to be considered as secondary reactions to more primary pressures from the side of the dominant white majority. . . .

. . . The Negro, as a minority, and a poor and suppressed minority at that, in the final analysis, has had little other strategy open to him than to play on the conflicting values held in the white majority group. In so doing, he has been able to identify his cause with broader issues in American politics and social life and with moral principles held dear by the white Americans. This is the situation even today and will remain so in the foreseeable future. In that sense, "this is a white man's country."

---

[1]This is less true, of course, in communities where the ratio between the number of Negroes and the number of whites diverges sharply from the average ratio of one to ten for the whole nation.

This stress in the formulation of our problem, it must be repeated, is motivated by an ambition to be realistic about the actual power relations in American society. It should not be taken as a doctrinaire approach. In the degree that the Negro people succeed in acquiring and institutionalizing footholds of power in society with the help of interested white groups—for example, if they can freely use their votes, as they can in the North, or press themselves into the industrial labor market and the trade unions—they will increasingly be able to act and not only to react. Under all circumstances, in fact even in slavery, the attitudes and activities of the Negro people do, to a certain extent, influence the attitudes and policies of the white majority group in power, as account is taken by the whites of the Negro's reactions. Even if the prevailing power situation is reason enough to look for the primary responsibility for what happens in the valuations of the white people, these same valuations are themselves the product of a two-way interracial relationship.

Not an Isolated Problem

Closely related to the thesis that the Negro problem is predominantly a white man's problem is another conclusion, which slowly dawned upon the author, though it undoubtedly is not news to many of his American readers: *The Negro problem is an integral part of, or a special phase of, the whole complex of problems in the larger American civilization. It cannot be treated in isolation.* There is no single side of the Negro problem—whether it be the Negro's political status, the education he gets, his place in the labor market, his cultural and personality traits, or anything else—which is not predominantly determined by its total American setting. We shall, therefore, constantly be studying the American civilization in its entirety, though viewed in its implications for the most disadvantaged population group.

There is a natural tendency on the part of white people in America to attempt to localize and demarcate the Negro problem into the segregated sector of American society where the Negroes live. This tendency is visible even in many scientific treatments of the Negro problem. The Negro spokesmen, on their side, are often equally tempted to stress the singularity of their grievances to the extent of not considering the broader setting. The fact of segregation also often makes them less familiar with the American society at large. The Negro social scientists have their special opportunity in knowing intimately the Negro community and will—with a few outstanding exceptions—treat their problems in isolation. . . .

The relationship between American society and the Negro problem is not one-sided. The entire structure of American society is itself greatly conditioned by the presence of the thirteen million Negro citizens. American politics, the labor market, education, religious life, civic ideals, art, and recreation are as they are partly because of the important conditioning factor working throughout the history of the nation. New impulses from the Negro people are constantly affecting the American way of life, bending in some degree all American institutions and bringing changes in every aspect of the American's complex world view. While primary attention will be focused on the Negro people and on the influences *from* the larger society working on them, their influence *back on* white society will not be ignored. . . .

America's Opportunity

. . . In the present phase of history this is what the world needs to believe. Mankind is sick of fear and disbelief, of pessimism and cynicism. It needs the youthful moralistic optimism of America. But empty declarations only deepen cynicism. Deeds are called for. If America in actual practice could show the world a progressive trend by which the Negro became finally integrated into modern democracy, all mankind would be given faith again—it would have reason to believe that peace, progress and order are feasible. And America would have a spiritual power many times stronger than all her financial and military resources—the power of the trust and support of all good people on earth. *America is free to choose whether the Negro shall remain her liability or become her opportunity.*

The development of the American Negro problem during the years to come is, therefore, fateful not only for America itself but for all mankind. If America wants to make the second choice, she cannot wait and see. She has to do something big and do it soon. . . .

We are now in a deeply unbalanced world situation. Many human relations will be readjusted in the present world revolution, and among them race relations are bound to change considerably. As always in a revolutionary situation when society's moorings are temporarily loosened, there is, on the one hand, an opportunity to direct the changes into organized reforms and, on the other hand, a corresponding risk involved in letting the changes remain uncontrolled and lead into disorganization. To do nothing is to accept defeat.

From the point of view of social science, this means, among other things, that social engineering will increasingly be demanded. Many things that for a long period have been predominantly a matter of individual adjustment will become more and more determined by political decision and public regulation. We are entering an era where fact-finding and scientific theories of causal relations will be seen as instrumental in planning controlled social change. The peace will bring nothing but problems, one mounting upon another, and consequently, new urgent tasks for social engineering. The American social scientist, because of the New Deal and the War, is already acquiring familiarity with planning and practical action. He will never again be given the opportunity to build up so "disinterested" a social science.

The social sciences in America are equipped to meet the demands of the post-war world. In social engineering they will retain the old American faith in human beings which is all the time becoming fortified by research as the trend continues toward environmentalism in the search for social causation. In a sense, the social engineering of the coming epoch will be nothing but the drawing of practical conclusions from the teaching of social science that "human nature" is changeable and that human deficiencies and unhappiness are, in large degree, preventable.

In this spirit, so intrinsically in harmony with the great tradition of the Enlightenment and the American Revolution, the author may be allowed to close with a personal note. Studying human beings and their behavior is not discouraging. When the author recalls the long gallery of persons whom, in the course of this inquiry, he has come to know with the impetuous but temporary intimacy of the stranger — sharecroppers and plantation owners, workers and employers, merchants and bankers, intellectuals, preachers, organization leaders, political bosses, gangsters, black and white, men and women, young and old, Southerners and Northerners — the general observation retained is the following: Behind all outward dissimilarities, behind their contradictory valuations, rationalizations, vested interests, group allegiances and animosities, behind fears and defense constructions, behind the role they play in life and the mask they wear, people are all much alike on a fundamental level. And they are all good people. They want to be rational and just. They all plead to their conscience that they meant well even when things went wrong.

Social study is concerned with explaining why all these potentially and intentionally good people so often make life a hell for themselves and each other when they live together, whether in a family, a community, a nation or a world. The fault is certainly not with becoming organized *per se*. In their formal organizations, as we have seen, people invest their highest ideals. These institutions regularly direct the individual toward more cooperation and justice than he would be inclined to observe as an isolated private person. The fault is, rather, that our structures of organizations are too imperfect, each by itself, and badly integrated into a social whole.

The Petitioner's Brief in
Sweatt v. Painter

1950

As Plessy v. Ferguson *validated the increasingly common southern practice of separate but equal public arrangements for blacks and whites, it legally sustained grossly unequal schools. In decisions like* Berea College v. Kentucky *(1908), the Supreme Court upheld the constitutionality of Jim Crow education. Beginning in the 1930s and with moderate success, the legal arm of the NAACP mounted a series of challenges to Jim Crow in graduate and professional education. For example, in* Gaines ex rel. Canada v. Missouri *(1938) and* Sipuel v. Oklahoma State Regents *(1948), the Court ruled that the state's failure to provide a legal education for blacks was a violation of their Fourteenth Amendment equal rights guarantee.*

Two 1950 cases, Sweatt v. Painter *and* McLaurin v. Oklahoma State Regents, *were critical to the expanding legal struggle against Jim Crow education. Following these decisions, states could no longer provide unequal arrangements for blacks. In addition, it had been common practice for various southern states to pay for black students to enroll in professional and graduate programs outside their states, typically in the North, rather than allow them to attend all-white schools in the state or region. Under the more stringent guidelines necessitated by* Sweatt *and* McLaurin, *separate schools had to be truly equal. Such a standard was financially and practically unattainable in the postwar South.*

Another victory in both Sweatt *and* McLaurin *was the mandated remedy. The Court argued that the relief for the claimants had to be immediate because the rights infringed upon were "present and personal." Thus there could be no delay in admitting the black applicants to the state schools of their choice. The thrust of the high court ruling in each case was clearly against the grain of* Plessy. *It was evident, however, that the Court was unprepared to take the next step, which these plaintiffs vigorously argued for, and overturn* Plessy.

In 1946 Heman Sweatt, a black letter carrier, sought admission to the University of Texas School of Law. After the university denied his application on the basis of race rather than his credentials, Sweatt filed suit against the university claiming that his right to equal educational opportunity had been denied. The state's solution was to create a separate and substantially

---

Sweatt v. Painter, 339 U.S. 629 (1950).

equal law school for blacks in Houston. Until that school could be realized,
a separate, makeshift law school for blacks would be established in Austin
in affiliation with the University of Texas. The NAACP had helped shape
Sweatt's initial challenge and argued his case all the way to the Supreme
Court, where Sweatt finally prevailed.

Sweatt's brief reprinted here, turned on two key arguments, the substance
of which is traceable in previous and subsequent legal argumentation push-
ing integration and the abolition of Jim Crow. First, the patently unequal
solution of the makeshift black school exposed "separate but equal" as the fic-
tion it was and violated Sweatt's equal rights. Second, according to the brief,
the case was sufficiently strong to serve as a basis for overturning the prece-
dent of Plessy. The United States had reached the point in its race relations
and judicial history where it had to articulate an integrationist as opposed
to a segregationist legal ethic. Also, equal educational opportunity served com-
pelling public concerns, including the need for many more black leaders in
concert with the need for increased numbers of blacks with graduate and pro-
fessional training. The brief furthermore argued that segregation harmed
whites and blacks, while integration would advance the best interests of both
groups. It is useful to assess the cogency of the legal arguments. Does the brief
persuasively show integration's alleged positive effects? How do the integra-
tionist positions here compare with those in Sumner's brief in Roberts v.
Boston *(p. 57) and Harlan's dissent in* Plessy v. Ferguson *(p. 81)?*

Argument

I

The question whether a state which undertakes to provide legal
education for any of its citizens can satisfy the requirements of
the equal protection clause of the Fourteenth Amendment by
establishing a law school for Negroes separate from the law school
it provides for all other persons is of great public importance and
should be decided by this Court in this case.

The education of the youth of our nation, formerly the responsibility
of the parent, has now become a recognized function of government. This
has become a matter of national importance. The individual states have
provided public education through the graduate and professional school
levels. Most of the states provide educational facilities without regard to
the race or creed of the student. However, seventeen of the states have
insisted upon either the complete exclusion or the segregation of

Negroes in public education.[1] The record of these states has brought down the national level of education. The question of the legality of such racial segregation, which amounts to actual exclusion from the regular recognized state university, is of great public importance. . . .

Recognizing that segregation constitutes a menace to American freedom and was indefensible, the President's Committee on Civil Rights unequivocally recommended its elimination from American life.[2] In the same year, the President's Commission on Higher Education, in its report on education in the United States said:[3]

> The time has come to make public education at all levels equally accessible to all, without regard to race, creed, sex or national origin.

This, too, is the almost unanimous conclusion of scholars and students who have studied the problem.

The professional skills developed through graduate training are among the most important elements of our society. . . .

Racial inequality in education has resulted in a loss to the nation of the development of these professional skills in a great part of our population. Because of the limited opportunities open to Negroes in professional education, in the United States in 1940, there was one white physician for every 735 white citizens, but only one Negro doctor for every 3,651 Negroes.[4] And one white lawyer served 670 whites, but there was only one colored lawyer for every 12,230 Negro citizens.[5] In the petitioner's native state of Texas, the same deprivation of professional services exists. In 1940 in Texas, one white lawyer served 709 whites, whereas there was only one Negro lawyer for every 40,191 Negroes.[6]

---

[1]Alabama, Arkansas, Delaware, Florida, Georgia, Kentucky, Louisiana, Maryland, Mississippi, Missouri, North Carolina, Oklahoma, South Carolina, Tennessee, Texas, Virginia, West Virginia.

[2]"To Secure These Rights," *The Report of the President's Committee on Civil Rights* (Washington, DC: U.S. Government Printing Office, 1947) 166. "The separate but equal doctrine has failed in three important respects. First, it is inconsistent with the fundamental equalitarianism of the American way of life in that it marks groups with the brand of inferior status. Secondly, where it has been followed, the results have been separate and unequal facilities for minority peoples. Finally, it has kept people apart despite incontrovertible evidence that an environment favorable to civil rights is fostered whenever groups are permitted to live and work together. There is no adequate defense of segregation." Ibid.

[3]"Higher Education for American Democracy," *A Report of the President's Commission on Higher Education* (Washington, DC: U.S. Government Printing Office, December 1947) 38.

[4]*Journal of Negro Education,* Vol. XIV (Fall 1945), 511.

[5]Ibid., 512.

[6]Based on data in *Sixteenth Census of the United States:* Population, Vol. III, Part 4, Reports by States (1940).

Perhaps even more important than the barriers which segregation offers to the development of leadership and professional skills is its corrosive effect upon the fundamentals of a democratic society. Neither white nor Negro Americans can maintain complete and full allegiance to the basic tenet upon which our government is founded — "that all men are created equal"—when pupils are being forcibly kept apart in the public schools because of their racial identity.

II

The inconsistency between the judicial approval of laws imposing racial distinctions in *Plessy* v. *Ferguson* and the judicial disapproval of similar distinctions and classifications in more recent decisions should lead this Court to review and disavow the doctrine of *Plessy* v. *Ferguson*.

. . . *Plessy* v. *Ferguson* raised in this Court for the first time the question of the constitutionality of a state statute enforcing segregation based upon race and color. In that case, a Louisiana statute requiring the separation of Negro and white passengers was held to be consistent with the equal protection clause of the Fourteenth Amendment. Yet the opinion appears to rely heavily upon the leading state case in this field — and the only one of the cited cases discussed in the majority opinion[7]—*Roberts* v. *Boston,* 5 Cush. (Mass.) 198 (1849), decided almost twenty years before the adoption of the Fourteenth Amendment. Yet, it was the very diversity of opinion, so pronounced in 1849, on the reasonableness of legal distinctions based on race which the Fourteenth Amendment sought to settle. Antebellum justifications of segregation have no more logical place in the interpretation of the Fourteenth Amendment than antebellum notions of voting restrictions have in defining the scope and meaning of the Fifteenth Amendment.

In addition, *Plessy* v. *Ferguson* was decided upon pleadings which assumed a theoretical equality within segregation rather than on a full hearing and evidence which would have revealed equality to be impossible under a system of segregation.

An examination of the other decisions of this Court upon which the lower court relied shows that the doctrine of *Plessy* v. *Ferguson* has not been reexamined nor seriously challenged.

. . . The inconsistencies between the "separate but equal" doctrine of

---

[7]Other cases cited in the opinion include: *People* v. *Gallagher,* 93 N.Y. 438; and *Ward* v. *Flood,* 48 Cal. 36; *State, Barnes* v. *McCann,* 21 Ohio. St. 210; *Lehew* v. *Brummell,* 103 Mo. 546; *Cory* v. *Carter,* 48 Ind. 337; *Dawson* v. *Lee,* 83 Ky. 49.

Plessy v. *Ferguson* and the reasoning and holdings of a considerable body of decisions of this Court become readily apparent when analysis is made in terms of the fundamental question, common to all, whether racial differences can be made the bases for legislative distinctions in the face of the Fourteenth Amendment. Except in *Plessy* v. *Ferguson* . . . and the decisions which rely uncritically upon it, this Court has consistently concluded that the Fourteenth Amendment prohibits the states from making racial differences and other arbitrary distinctions the bases for general classifications. . . .

These cases merely apply to racial distinctions the general constitutional principle applicable in all other areas. Their rationale is merely a part of and consistent with the basic principle that all governmental classifications must be based upon a significant difference having a reasonable relationship to the subject matter of the statute. . . .

The Court . . . in relying on *Plessy* v. *Ferguson* and in ignoring this body of cases has improperly and mistakenly construed the limitations of the Fourteenth Amendment as applied to the instant case. . . . In requiring that a classification be based upon a significant difference having a reasonable relationship to the subject matter of the statute, that body of decisions rests upon a sound foundation. The same principle should be controlling in the instant case. Any other approach makes the equal protection clause meaningless. Insofar as *Plessy* v. *Ferguson* affects the application of that principle to the instant case, it should not be followed.

III

This Court should review and reverse the judgment below to prevent the several states from being free to restrict Negroes to public educational facilities clearly inferior to those provided for all other persons similarly situated through the device of arbitrary judicial decision that such discriminatory action provides "substantial equality."

Texas and sixteen other states have insisted that public education can only be furnished on a basis of racial distinction between students. The purpose of this practice is to exclude Negroes from the recognized state educational institutions. The record in this case, as in other cases, will demonstrate that these states first establish facilities for non-Negroes. Later, either as a result of legal action or other compulsion, separate institutions have been established for Negroes.

The record in this case, the record in similar cases, governmental and private studies, demonstrate clearly that the separate Negro facilities are never equal to the facilities established for other groups. In short, we have been unable to find a single recognized study of public education on a segregated basis which reveals equality of opportunity as between the segregated and now segregated [sic] schools. The Negro school is invariably an inferior school.

The "separate but equal" doctrine of *Plessy* v. *Ferguson,* relied upon by Texas and the other southern states, is based on the hypothesis that equal facilities can be realized in a segregated school system. The record in this case and in other cases has demonstrated the invalidity of such a hypothesis. It is clear not only that the doctrine of "separate but equal" has not produced equality, but can never provide the equality required by the Fourteenth Amendment.

This separate but equal doctrine has brought about constant and continual litigation. Negroes have gone to the courts in Missouri,[8] South Carolina,[9] Tennessee,[10] Louisiana,[11] Oklahoma,[12] Maryland,[13] Kentucky,[14] and Texas in order to secure educational advantages equal to those being offered to all other qualified persons. The formula constantly invites such court action. In all instances this has meant loss of time and years out of an individual's career, while his case pursues its way through the courts. This very fact shows the weakness of the doctrine.

. . . The most authoritative studies made on public education in the United States clearly indicate that the Negro institutions are vastly inferior to the whites. Yet when faced with the necessity of holding Negro institutions to be inferior to the white and therefore to order the admission of the Negro to the white institution, courts have fallen back on the formula "substantially equivalent" to justify their decision to refuse the admission of the Negro into the white institution. . . .

Therefore, the only way for the petitioner in this case and other qualified Negroes to obtain a legal education equal to that obtained by all other qualified applicants is by admission to the recognized state institutions. The only way this can be accomplished is for this Court to reconsider the doctrine of *Plessy* v. *Ferguson* and overrule it.

[8]*Missouri, ex rel. Gaines* v. *Canada, supra. Bluford* v. *Canada,* 32 F. Supp. 707 (1940) (Appeal dismissed 119 F. (2d) 779 (C.C.A. 8th)).

[9]*Wrighten* v. *Board of Trustees,* 72 F. Supp. 948.

[10]*State, ex rel. Michael* v. *Whitman,* 179 Tenn. 250, 165 S.W. (2d) 378 (1942).

[11]*Louisiana, ex rel. Hatfield* v. *Louisiana State University* (File 25,520, State Court for the 19th Judicial District).

[12]*Sipuel* v. *Board of Regents, supra; Fisher* v. *Hurst, supra; McLaurin* v. *Oklahoma State Regents,* No. 614, U.S. Supreme Court, Oct. Term, 1948.

[13]*Pearson* v. *Murray,* 169 Md. 478, 182 Atl. 590 (1936).

[14]*Johnson* v. *Board of Trustees* (File No. 625, U.S. Dist. Court for the Eastern Dist. of Kentucky).

CHIEF JUSTICE FRED VINSON

Opinion of the Court in Sweatt v. Painter

1950

The Court decided that Texas's efforts to create a substantially equal but separate law school for blacks were grossly inadequate and thus created an unequal education. Interestingly, Chief Justice Vinson's majority opinion advanced the argument that a law school's quality was determined by a variety of criteria, including the school's reputation, the professional connections it offered, and the eminence of its faculty. Using these measures, Texas's forthcoming black law school could not possibly match its all-white law school anytime soon. This denial of the plaintiff's right to equal educational opportunity, this infringement of his "personal and present rights," demanded immediate relief—his admission without delay to the University of Texas Law School.

Although the decision favored the plaintiff, the justices still refused to strike down Plessy. *Nevertheless, the opinion inspired hope as it strongly suggested that a high court ruling against* Plessy *was both plausible and imminent. How do you explain the Court's position? Do you think that Vinson's criteria signify an accurate measure of a law school's quality? What do you see as the critical similarities and differences between Jim Crow education at the elementary and high school levels as opposed to the graduate and professional levels? Why was it easier to battle the latter? Why was Sweatt successful and Roberts and Plessy were not?*

. . . This case and *McLaurin* v. *Oklahoma State Regents* . . . present different aspects of this general question: To what extent does the Equal Protection Clause of the Fourteenth Amendment limit the power of a state to distinguish between students of different races in professional and graduate education in a state university? . . .

Whether the University of Texas Law School is compared with the original or the new law school for Negroes, we cannot find substantial equality in the educational opportunities offered white and Negro law students by the State. In terms of number of the faculty, variety of courses and opportunity for specialization, size of the student body, scope of the library, availability of law review and similar activities, the University of Texas Law School

---

Sweatt v. Painter, 339 U.S. 629 (1950).

is superior. What is more important, the University of Texas Law School possesses to a far greater degree those qualities which are incapable of objective measurement but which make for greatness in a law school. Such qualities, to name but a few, include reputation of the faculty, experience of the administration, position and influence of the alumni, standing in the community, traditions and prestige. It is difficult to believe that one who had a free choice between these law schools would consider the question close.

Moreover, although the law is a highly learned profession, we are well aware that it is an intensely practical one. The law school, the proving ground for legal learning and practice, cannot be effective in isolation from the individuals and institutions with which the law interacts. Few students and no one who has practiced law would choose to study in an academic vacuum, removed from the interplay of ideas and the exchange of views with which the law is concerned. The law school to which Texas is willing to admit petitioner excludes from its student body members of the racial groups which number 85% of the population of the State and include most of the lawyers, witnesses, jurors, judges and other officials with whom petitioner will inevitably be dealing when he becomes a member of the Texas Bar. With such a substantial and significant segment of society excluded, we cannot conclude that the education offered petitioner is substantially equal to that which he would receive if admitted to the University of Texas Law School.

It may be argued that excluding petitioner from that school is no different from excluding white students from the new law school. This contention overlooks realities. It is unlikely that a member of a group so decisively in the majority, attending a school with rich traditions and prestige which only a history of consistently maintained excellence could command, would claim that the opportunities afforded him for legal education were unequal to those held open to petitioner. That such a claim, if made, would be dishonored by the State, is no answer. . . .

It is fundamental that these cases concern rights which are personal and present. This Court has stated unanimously that "The State must provide [legal education] for [petitioner] in conformity with the equal protection clause of the Fourteenth Amendment and provide it as soon as it does for applicants of any other group." *Sipuel* v. *Board of Regents* . . . (1948). That case "did not present the issue whether a state might not satisfy the equal protection clause of the Fourteenth Amendment by establishing a separate law school for Negroes." *Fisher* v. *Hurst* . . . (1948). In *Missouri ex rel. Gaines* v. *Canada* . . . (1938), the Court, speaking through Chief Justice Hughes, declared that "petitioner's right was a personal one. It was as an individual that he was entitled to the equal protection of the laws, and the State was bound to furnish him within its borders facilities for legal education substantially equal to those which the State there afforded for persons of the white race, whether or not other negroes sought the same opportunity." These are the only cases in this Court which present the issue of the constitutional validity of race distinctions in state-supported graduate and professional education.

In accordance with these cases, petitioner may claim his full constitutional right: legal education equivalent to that offered by the State to students of other races. Such education is not available to him in a separate law school as offered by the State. . . .

We hold that the Equal Protection Clause of the Fourteenth Amendment requires that petitioner be admitted to the University of Texas Law School. The judgment is reversed and the cause is remanded for proceedings not inconsistent with this opinion.

CHIEF JUSTICE FRED VINSON

Opinion of the Court in
McLaurin v. Oklahoma State Regents

1950

George McLaurin, a black schoolteacher in his sixties, sought admission to the University of Oklahoma's School of Education in 1948 to advance his education. Because the state had made no provisions, separate or otherwise, for black graduate education, McLaurin was admitted to the program at Oklahoma, but under a variety of Jim Crow arrangements, including having to sit in a room off to the side of the regular classroom. The NAACP fought this denial of equal educational opportunity all the way to the Supreme Court.

As in Sweatt, *Chief Justice Vinson's opinion accepted as fact that the plaintiff's education was unequal under the state's apartheid restrictions and that this was a violation of his right to equal educational opportunity. The rule of equality thus necessarily extended to the treatment of blacks within institutions defined as all-white. Indeed Vinson advanced the intriguing argument that the separate accommodations for McLaurin were a handicap. Again, as in* Sweatt, *the Court ordered immediate relief—access to a truly equal education.*

McLaurin v. *Oklahoma State Regents for Higher Education et al.,* 339 U.S. 637 (1950).

While refusing to go the next step and overturn Plessy, *this decision, like that in* Sweatt, *boded ill for Jim Crow's principal legal sanction. Ultimately, the decision to overrule* Plessy *demanded a measure of courage and political savvy that eluded the high court under Vinson's leadership. Nevertheless, what do you see as the strengths and weaknesses of Vinson's decision? Why is it useful to compare and contrast the decisions in* Sweatt *and* McLaurin? *Are there salient differences that parallel the explicit similarities in both cases? If so, delineate and analyze them. Finally, how did* McLaurin, *like* Sweatt, *pave the way for* Brown?

. . . It is said that the separations imposed by the State in this case are in form merely nominal. McLaurin uses the same classroom, library and cafeteria as students of other races; there is no indication that the seats to which he is assigned in these rooms have any disadvantage of location. He may wait in line in the cafeteria and there stand and talk with his fellow students, but while he eats he must remain apart.

These restrictions were obviously imposed in order to comply, as nearly as could be, with the statutory requirements of Oklahoma. But they signify that the State in administering the facilities it affords for professional and graduate study, sets McLaurin apart from the other students. The result is that appellant is handicapped in his pursuit of effective graduate instruction. Such restrictions impair and inhibit his ability to study, to engage in discussions and exchange views with other students, and, in general, to learn his profession.

Our society grows increasingly complex, and our need for trained leaders increases correspondingly. Appellant's case represents, perhaps, the epitome of that need, for he is attempting to obtain an advanced degree in education, to become, by definition, a leader and trainer of others. Those who will come under his guidance and influence must be directly affected by the education he receives. Their own education and development will necessarily suffer to the extent that his training is unequal to that of his classmates. State-imposed restrictions which produce such inequalities cannot be sustained.

It may be argued that appellant will be in no better position when these restrictions are removed, for he may still be set apart by his fellow students. This we think irrelevant. There is a vast difference—a Constitutional difference—between restrictions imposed by the state which prohibit the intellectual commingling of students, and the refusal of individuals to commingle where the state presents no such bar. . . . The removal of the state restrictions will not necessarily abate individual and group predilections, prejudices and choices. But at the very least, the state will not be depriving appellant of the opportunity to secure acceptance by his fellow students on his own merits.

We conclude that the conditions under which this appellant is required to receive his education deprive him of his personal and present right to the equal protection of the laws. . . . We hold that under these circumstances the Fourteenth Amendment precludes difference in treatment by the state based upon race. Appellant, having been admitted to a state-supported graduate school, must receive the same treatment at the hands of the state as students of other races. The judgment is

Reversed.

4

Brown v. Board of Education

(1952–55)

The central issue in *Brown v. Board of Education* was the constitutionality of segregated elementary and secondary schools. Could separate actually be equal? Did these typically separate and unequal schools violate the constitutional right of blacks to equality before the law as propounded in the Fourteenth Amendment? Finally, should the legal rationale for Jim Crow—*Plessy*—be overturned?

The high court ruling came in two parts. *Brown I* was the original decision on the merits of the case. *Brown II* was the decision regarding remedy or relief for the plaintiffs. The documents in this chapter reflect the structure of the case itself. The first set of documents presents a sample of the pre–Supreme Court proceedings—the trial court, or lower court, stage. (The Supreme Court most often rules on cases appealed from the lower courts.) The lower court selections here come from *Briggs v. Elliott*, a case that vividly encapsulates the fundamental contours and meanings of the overall defense and plaintiff positions in all of the cases constituting *Brown*. Included here are the majority opinion in favor of Clarendon County's segregated school system as well as a dissent arguing against both that system and *Plessy* itself.

The actual Supreme Court deliberations are divided into three sections, or rounds. While the actual oral arguments before the Court are indeed important, these often highly dramatic rhetorical moments—unlike the formally prepared documents—lack the formal narrative features of organization, development, internal coherence, and persuasive power. As a result, they are not included. Instead, each section cites the most important pieces of evidence from the written record.

Round one looks at the first stage of litigation before the Supreme Court in late 1952. This round of documents lays out both sides of the case, building upon the substance and form of their previous positions and foreshadowing the substance and form of their subsequent ones.

In round two, deliberations narrowed to a particular set of questions that

The bottom portrait of a black school in Autauga County, Alabama, captures an early 1920s replacement for the one pictured above it. The rebuilt school (with aid from the Rosenwald Fund) was a proud moment for the local black community. Visually, it epitomizes a resourceful commitment to racial advancement.

the justices want both sides to answer. In this setting, the issue was to establish the original intent of the legislative framers of the Fourteenth Amendment. Did they see the amendment as supporting or opposing segregated schools? With this knowledge, the judges could weigh judicial precedence and tradition more effectively. The justices also asked both sides for their thoughts about the kind of remedy they envisioned in the case.

Round three focuses on what turned out to be the hardest question of all: what kind of relief to hand down. The three documents in this section are pivotal to an understanding of this perplexing problem. First, in their brief the successful plaintiffs intensified their demand for immediate relief. Second, the defeated defendants asked for incremental relief. Last, Chief Justice Warren's official opinion decreed a remedy of "all deliberate speed."

Mississippi Voter Registration Form

1955

In the late nineteenth and early twentieth centuries, the often hard-fought white campaigns to expel black voters from the southern electorate were a critical component of the institutionalization of white supremacy. Notwithstanding opposition from blacks and their allies these disfranchisement offensives succeeded. Legal and illegal tactics such as intimidation, violence, and murder marked these campaigns. As a result, by the early twentieth century, the number of black voters throughout the South had dropped drastically; in many communities, the black vote was more or less wiped out.

In fact, the various "legal" disfranchisement devices employed over time— especially literacy tests and poll taxes—significantly diminished the size of the poor and working-class white sector of the southern electorate. Blacks, however, bore the brunt of disfranchisement. Widespread illiteracy made literacy tests — requirements that voters "read" a section of the state or federal constitution — quite effective. Likewise, poll taxes — annual head taxes averaging around three dollars annually — disfranchised vast numbers because of widespread poverty. Nevertheless, it should be kept in mind that these offensives to strip blacks of the vote had a class as well as racial impact, further entrenching the power of relatively small white southern political elites, or machines. Nowhere was this clearer than in the remaking of the southern Democratic Party into a series of private white clubs dominated by the "better sort" who controlled southern politics. This systematic exclusion of blacks from state Democratic Party primary elections — known as the white primary — allowed white southern congressmen to parlay seniority and control over a small white constituency into enormous power at the national level.

The battle to restore the black vote, therefore, was vital to the ongoing black freedom struggle. In 1944, the NAACP achieved a major Supreme Court victory in Smith v. Allwright, *which outlawed the white primary.*

Whereas previous rulings interpreted the white primary as consistent with the Fourteenth and Fifteenth Amendments because of the Democratic Party's status as a private organization, Smith v. Allwright *ruled against this legal fiction of state action masquerading as private action.*

The Mississippi Voter Registration application reprinted here was part of a series of devices that effectively disfranchised blacks as late as the mid-1960s because satisfactory completion of the form was at the discretion of the white registrar, who reviewed the forms of whites far less strictly. Such byzantine writing hurdles worked in concert with irregular voting hours; requiring registration at the county courthouse—the center of local white power—rather than local precincts; and harassment, violence, and even murder to discourage black registration. On the voter registration form, any error—real or imagined, large or small—was typically sufficient for the white registrar to deny a black applicant. Particularly effective was the question demanding that the applicant explain a section of the Mississippi state constitution to the registrar's satisfaction. Similar "understanding" requirements and disfranchisement devices were indeed common throughout the South.

Various southern black voter registration campaigns within the Civil Rights Movement fought to undo and surmount these restrictions. Consequently, the Voting Rights Act of 1965, which made age and proof of residency the only binding voting requirements, constituted a powerful victory for the struggle. In turn, the Voting Rights Act soon helped usher in the extraordinary post-1965 growth in the black southern electorate, the equally extraordinary expansion in the numbers of southern black elected officials, and, more broadly, the phenomenal rise of black political power in the South.

Does anything about this form strike you as noteworthy or unusual? In what ways is it race-neutral? In what ways is it race-conscious? Do you imagine that you could have successfully completed such an application at the time?

Sworn Written Application for Registration

(By reason of the provisions of Section 244 of the Constitution of Mississippi and House Bill No. 95, approved March 24, 1955, the applicant for registration, if not physically disabled, is required to fill in this form in his own handwriting in the presence of the registrar and without assistance or suggestion of any other person or memorandum)

1. Write the date of this application: _____
2. What is your full name?_____
3. State your age and date of birth:_____

4. What is your occupation? _____

5. Where is your business carried on? _____

6. By whom are you employed? _____

7. Are you a citizen of the United States and an inhabitant of Mississippi?_____

8. For how long have you resided in Mississippi? _____

9. Where is your place of residence in the district?_____

10. Specify the date when such residence began: _____

11. State your prior place of residence, if any:_____

12. Check which oath you desire to take: (1) General _____ (2) Minister's _____ (3) Minister's Wife _____ (4) If under 21 years at present, but 21 years by date of general election _____

13. If there is more than one person of your same name in the precinct, by what name do you wish to be called? _____

14. Have you ever been convicted of any of the following crimes: bribery, theft, arson, obtaining money or goods under false pretenses, perjury, forgery, embezzlement, or bigamy? _____

15. If your answer to Question 14 is "Yes", name the crime or crimes of which you have been convicted, and the date and place of such conviction or convictions:_____

16. Are you a minister of the gospel in charge of an organized church, or the wife of such a minister? _____

17. If your answer to Question 16 is "Yes", state the length of your residence in the election district _____

18. Write and copy in the space below, Section _____ of the Constitution of Mississippi.

(Instruction to Registrar: You will designate the section of the Constitution and point out same to applicant)

19. Write in the space below a reasonable interpretation (the meaning) of the section of the Constitution of Mississippi which you have just copied:

[Writing space on form]

20. Write in the space below a statement setting forth your understanding of the duties and obligations of citizenship under a constitutional form of government.

[Writing space on form]

21. Sign and attach hereto the oath or affirmation named in Question 12.

[Writing space on form]

_____

The applicant will sign his name here.

STATE OF MISSISSIPPI,
COUNTY OF _____
Sworn to and subscribed before me by the within named _____
on this the _____ day of _____ 19_____.

_____

COUNTY REGISTRAR

THE LOWER COURT ROUND: PRELIMINARY DELIBERATIONS

JUDGE JOHN J. PARKER

Decision in Briggs v. Elliott
1951

Briggs v. Elliott, *a South Carolina case decided in 1951 by a three-judge District Court, graphically captured the contending positions in the various cases that would soon be argued together as the* Brown *litigation. The plaintiffs in* Briggs, *represented by Thurgood Marshall, the head of the NAACP legal team, argued two principal positions. First, they maintained that the inferiority of separate black schools demonstrated that separate was inherently unequal and a violation of the plaintiffs' right to equal educational opportunity. It thus followed that rather than reforming this inequitable and harmful system, the courts had to abolish Jim Crow schools at once in favor of integrated schools. Second, the plaintiffs contended that the precedent of* Plessy *had to be overturned.*

Briggs v. Elliott, 98 F. Supp. 529 (1951), 531–37.

The defendants, represented by eminent lawyer-politician John W. Davis, argued first that the separate and admittedly unequal black schools did not violate the equal rights of the black schoolchildren because these schools were part of a Jim Crow system that had the imprimatur of law as well as custom. As a result, they stressed that the system of Jim Crow was basically sound in theory but needed to be reformed in practice. In this instance, the black schools had to be upgraded, or equalized. The second thrust of the defense argument followed logically from the first: Plessy *should be sustained.*

Judge John J. Parker's majority opinion, with Judge George Timmerman concurring, called for gradual equalization of black and white schools and sustained Plessy. *Judge J. Waties Waring's dissent argued that segregation was inherently unequal and that* Plessy *was outdated and had to be overturned.*

Which argument do you find more compelling or convincing? Why? How does Parker argue that black schools be made equal? How does Waring counter this "solution?" How do you explain the NAACP's legal team's inability to prevail in this case?

. . . At the beginning of the hearing the defendants admitted upon the record that "the educational facilities, equipment, curricula and opportunities afforded in School District No. 22 for colored pupils . . . are not substantially equal to those afforded for white pupils." The evidence offered in the case fully sustains this admission. The defendants contend, however, that the district is one of the rural school districts . . . providing educational facilities for the children of either race, and that the inequalities have resulted from limited resources and from the disposition of the school officials to spend the limited funds available "for the most immediate demands rather than in the light of the overall picture." They state that under the leadership of Governor Byrnes the Legislature of South Carolina has made provision for a bond issue of $75,000,000 with a three per cent sales tax to support it for the purpose of equalizing educational opportunities and facilities throughout the state and of meeting the problem of providing equal educational opportunities for Negro children where this had not been done. They have offered evidence to show that this educational program is going forward and that under it the educational facilities in the district will be greatly improved for both races and that Negro children will be afforded educational facilities and opportunities in all respects equal to those afforded white children.

. . . There can be no question but that where separate schools are maintained for Negroes and whites, the educational facilities and opportunities afforded by them must be equal. The state may not deny to any person within its jurisdiction the equal protection of the laws, says the Fourteenth Amendment; and this means that, when the state undertakes public education, it may not discriminate against any individual on account of race but must offer equal opportunity to all. . . . We think it clear, therefore, that plaintiffs are entitled to a declaration to the effect that the school facilities now afforded Negro children in District No. 22 are not equal to the facilities afforded white children in the district and to a mandatory injunction requiring that equal facilities be afforded them. How this shall be done is a matter for the school authorities and not for the court, so long as it is done in good faith and equality of facilities is afforded; but it must be done promptly and the court in addition to issuing an injunction to that effect will retain the case upon its docket for further orders and will require that defendants file within six months a report showing the action that has been taken by them to carry out the order.

. . . Plaintiffs ask that, in addition to granting them relief on account of the inferiority of the educational facilities furnished them, we hold that segregation of the races in the public schools, as required by the Constitution and statutes of South Carolina, is of itself a denial of the equal protection of the laws guaranteed by the Fourteenth Amendment, and that we enjoin the enforcement of the constitutional provision and statute requiring it and by our injunction require defendants to admit Negroes to schools to which white students are admitted within the district. We think, however, that segregation of the races in the public schools, so long as equality of rights is preserved, is a matter of legislative policy for the several states, with which the federal courts are powerless to interfere.

. . . One of the great virtues of our constitutional system is that, while the federal government protects the fundamental rights of the individual, it leaves to the several states the solution of local problems. In a country with a great expanse of territory with peoples of widely differing customs and ideas, local self government in local matters is essential to the peace and happiness of the people in the several communities as well as to the strength and unity of the country as a whole. It is universally held, therefore, that each state shall determine for itself, subject to the observance of the fundamental rights and liberties guaranteed by the federal Constitution how it shall exercise the police power, i.e. the power to legislate with respect to the safety, morals, health and general welfare. And in no field is this right of the several states more clearly recognized than in that of public education. . . .

Plaintiffs rely upon expressions contained in opinions relating to professional education such as *Sweatt v. Painter,* . . . *McLaurin v. Oklahoma State Regents,* . . . where equality of opportunity was not afforded. *Sweatt v. Painter,* however, instead of helping them, emphasizes that the separate but equal doctrine of *Plessy v. Ferguson,* has not been overruled, since the Supreme Court, although urged to overrule it, expressly refused to do so and based its decision on the ground that the educational facilities offered Negro law students in that case were not equal to those offered white students. . . . The case of *McLaurin v. Oklahoma State Regents,* involved humiliating and embarrassing treatment of a Negro graduate student to which no one should have been required to submit. Nothing of the sort is involved here.

The problem of segregation as applied to graduate and professional education is essentially different from that involved in segregation in education at the lower levels. In the graduate and professional schools the problem is one of affording equal educational facilities to persons sui juris[1] and of mature personality. Because of the great expense of such education and the importance of the professional contacts established while carrying on the educational process, it is difficult for the state to maintain segregated schools for Negroes in this field which will afford them opportunities for education and professional advancement equal to those afforded by the graduate and professional schools maintained for white persons. What the courts have said, and all they have said in the cases upon which plaintiffs rely is that, notwithstanding these difficulties, the opportunity afforded the Negro student must be equal to that afforded the white student and that the schools established for furnishing this instruction to white persons must be opened to Negroes if this is necessary to give them the equal opportunity which the Constitution requires.

The problem of segregation at the common school level is a very different one. At this level, as good education can be afforded in Negro schools as in white schools and the thought of establishing professional contacts does not enter into the picture. Moreover, education at this level is not a matter of voluntary choice on the part of the student but of compulsion by the state. The student is taken from the control of the family during school hours by compulsion of law and placed in control of the school, where he must associate with his fellow students. The law thus provides that the school shall supplement the work of the parent in the training of the child and in doing so it is entering a delicate field and one fraught with tensions and difficulties. In formulating educational policy at the common school level, therefore, the law must take account, not merely of the matter of affording instruction to the student, but also of the wishes of the parent as to the upbringing of the child and his associates in the formative period of childhood and adolescence. If public education

[1]Able to assume legal responsibility.

is to have the support of the people through their legislatures, it must not go contrary to what they deem for the best interests of their children.

There is testimony to the effect that mixed schools will give better education and a better understanding of the community in which the child is to live than segregated schools. There is testimony, on the other hand, that mixed schools will result in racial friction and tension and that the only practical way of conducting public education in South Carolina is with segregated schools. The questions thus presented are not questions of constitutional right but of legislative policy, which must be formulated, not in vacuo or with doctrinaire disregard of existing conditions, but in realistic approach to the situations to which it is to be applied. In some states, the legislatures may well decide that segregation in public schools should be abolished, in others that it should be maintained—all depending upon the relationships existing between the races and the tensions likely to be produced by an attempt to educate the children of the two races together in the same schools. The federal courts would be going far outside their constitutional function were they to attempt to prescribe educational policies for the states in such matters, however desirably such policies might be in the opinion of some sociologists or educators. For the federal courts to do so would result, not only in interference with local affairs by an agency of the federal government, but also in the substitution of the judicial for the legislative process in what is essentially a legislative matter. . . .

To this we may add that, when seventeen states and the Congress of the United States have for more than three-quarters of a century required segregation of the races in the public schools, and when this has received the approval of the Supreme Court of the United States at a time when that court included Chief Justice Taft and Justices Stone, Holmes and Brandeis, it is a late day to say that such segregation is violative of fundamental constitutional rights. It is hardly reasonable to suppose that legislative bodies over so wide a territory, including the Congress of the United States, and great judges of high courts have knowingly defied the Constitution for so long a period or that they have acted in ignorance of the meaning of its provisions. The constitutional principle is the same now that it has been throughout this period; and if conditions have changed so that segregation is no longer wise, this is a matter for the legislatures and not for the courts. The members of the judiciary have no more right to read their ideas of sociology into the Constitution than their ideas of economics.

. . . It is argued that, because the school facilities furnished Negroes in District No. 22 are inferior to those furnished white persons, we should enjoin segregation rather than direct the equalizing of conditions. In as much as we think that the law requiring segregation is valid, however, and that the inequality suffered by plaintiffs results, not from the law, but from the way it has been administered, we think that our injunction should be directed to removing the inequalities resulting from administration within the framework of the law rather than to nullifying the law itself. As a court of equity, we should exercise our power to assure to plaintiffs the equality of treatment to which they are entitled with due regard to the legislative policy of the state. In directing that the school facilities afforded Negroes within the district be equalized promptly with those afforded white persons, we are giving plaintiffs all the relief that they can reasonably ask and the relief that is ordinarily granted in cases of this sort. . . . The court should not use its power to abolish segregation in a state where it is required by law if the equality demanded by the Constitution can be attained otherwise. This much is demanded by the spirit of comity which must prevail in the relationship between the agencies of the federal government and the states if our constitutional system is to endure. . . .

JUDGE J. WATIES WARING

Dissent in Briggs v. Elliott

1951

. . . If a case of this magnitude can be turned aside and a court refused to hear these basic issues by the mere device of admission that some buildings, blackboards, lighting fixtures and toilet facilities are unequal but that they may be remedied by the spending of a few dollars, then, indeed people in the plight in which these plaintiffs are, have no adequate remedy or forum in which to air their wrongs. If this method of judicial evasion be adopted, these very infant plaintiffs now pupils in Clarendon County will probably be bringing suits for their children and grandchildren decades or rather generations hence in an effort to get for their descendants what are today denied to them. If they are entitled to any rights as American citizens, they are entitled to have these rights now. . . . And no excuse can be made to deny them these rights . . . by the false doctrine and patter called "separate but equal" and it is the duty of the Court to meet these issues simply and factually and without fear, sophistry and evasion. If this be the measure of justice to be meted out to them, then, indeed, hundreds, nay thousands, of cases will have to be brought and

Briggs v. Elliott, 98 F. Supp. 529 (1951), 538–48.

in each case thousands of dollars will have to be spent for the employ-ment of legal talent and scientific testimony and then the cases will be turned aside, postponed or eliminated by devices such as this.

We should be unwilling to straddle or avoid this issue and if the sug-gestion made by these defendants is to be adopted as the type of justice to be meted out by this Court, then I want no part of it.

And so we must and do face, without evasion or equivocation, the ques-tion as to whether segregation in education in our schools is legal or whether it cannot exist under our American system as particularly enunciated in the Fourteenth Amendment to the Constitution of the United States. . . .

Let us . . . consider whether the Constitution and Laws of the State of South Carolina . . . are in conflict with the true meaning and intendment of this Fourteenth Amendment. The whole discussion of race and ances-try has been intermingled with sophistry and prejudice. What possible definition can be found for the so-called white race, Negro race or other races? Who is to decide and what is the test? For years, there was much talk of blood and taint of blood. Science tells us that there are but four kinds of blood: A, B, AB and O and these are found in Europeans, Asiat-ics, Africans, Americans and others. And so we need to further consider the irresponsible and baseless references to preservation of "Caucasian blood." So then, what test are we going to use in opening our school doors and labeling them "white" and "Negro"? The law of South Carolina con-siders a person of one-eighth African ancestry to be a Negro. Why this proportion? Is it based upon any reason: anthropological, historical or eth-ical? And how are the trustees to know who are "whites" and who are "Negroes"? If it is dangerous and evil for a white child to be associated with another child, one of whose great-grandparents was of African descent, is it not equally dangerous for one with a one-sixteenth per-centage? And if the State has decided that there is a danger in contact between the whites and Negroes, isn't it requisite and proper that the State furnish a series of schools one for each of these percentages? If the idea is perfect racial equality in educational systems, why should children of pure African descent be brought in contact with children of one-half, one-fourth, or one-eighth such ancestry? To ask these questions is suffi-cient answer to them. The whole thing is unreasonable, unscientific and based upon unadulterated prejudice. We see the results of all of this warped thinking in the poor under-privileged and frightened attitude of so many of the Negroes in the southern states; and in the sadistic insis-tence of the "white supremacists" in declaring that their will must be imposed irrespective of rights of other citizens. This claim of "white supremacy," while fantastic and without foundation, is really believed by them for we have had repeated declarations from leading politicians and governors of this state and other states declaring that "white supremacy" will be endangered by the abolition of segregation. There are present threats, including those of the present Governor of this state, going to the extent of saying that all public education may be abandoned if the courts should grant true equality in educational facilities.

Although some 73 years have passed since the adoption of the Fourteenth Amendment and although it is clearly apparent that its chief purpose, (perhaps we may say its only real purpose) was to remove from Negroes the stigma and status of slavery and to confer upon them full rights as citizens, nevertheless, there has been a long and arduous course of litigation through the years. With some setbacks here and there, the courts have generally and progressively recognized the true meaning of the Fourteenth Amendment and have, from time to time, stricken down the attempts made by state governments (almost entirely those of the former Confederate states) to restrict the Amendment and to keep Negroes in a different classification so far as their rights and privileges as citizens are concerned. A number of cases have reached the Supreme Court of the United States wherein it became necessary for that tribunal to insist that Negroes be treated as citizens in the performance of jury duty . . . .

The Supreme Court has stricken down from time to time statutes providing for imprisonment for violation of contracts. These are known as peonage cases and were in regard to statutes primarily aimed at keeping the Negro "in his place."[1]

In the field of transportation the court has now, in effect declared that common carriers engaged in interstate travel must not and cannot segregate and discriminate against passengers by reason of their race or color.[2]

Frequent and repeated instances of prejudice in criminal cases because of the brutal treatment of defendants because of their color have been passed upon in a large number of cases.[3]

Discrimination by segregation of housing facilities and attempts to control the same by covenants have also been outlawed.[4]

---

[1] Peonage: Bailey v. Alabama, 219 U.S. 219, 31 S.Ct. 145, 55 L.Ed. 191; U.S. v. Reynolds, 235 U.S. 133, 35 S.Ct. 86, 59 L.Ed. 162.

[2] Transportation: Mitchell v. U.S., 313 U.S. 80, 61 S.Ct. 873, 85 L.Ed. 1201; Morgan v. Virginia, 328 U.S. 373, 66 S.Ct. 1050, 90 L.Ed. 1317; Henderson v. U.S., 339 U.S. 816, 70 S.Ct. 843, 94 L.Ed. 1302; Chance v. Lambeth, 4 Cir., 186 F.2d 879, certiorari denied Atlantic Coast Line R. Co. v. Chance, 341 U.S. 941, 71 S.Ct. 1001, May 28, 1951.

[3] Criminals: Brown v. Mississippi, 297 U.S. 278, 56 S.Ct. 461, 80 L.Ed. 682; Chambers v. Florida, 309 U.S. 227, 60 S.Ct. 472, 84 L.Ed. 716; Shepherd v. Florida, 341 U.S. 50, 71 S.Ct. 549.

[4] Housing: Buchanan v. Warley, 245 U.S. 60, 38 S.Ct. 16, 62 L.Ed. 149; Shelley v. Kraemer, 334 U.S. 1, 68 S.Ct. 836, 92 L.Ed. 1161.

In the field of labor employment and particularly the relation of labor unions to the racial problem, discrimination has again been forbidden.[5]

Perhaps the most serious battle for equality of rights has been in the field of exercise of suffrage. For years, certain of the southern states have attempted to prevent the Negro from taking part in elections by various devices. It is unnecessary to enumerate the long list of cases, but from time to time courts have stricken down all of these various devices classed as the "grandfather clause," educational tests and white private clubs.[6]

The foregoing are but a few brief references to some of the major landmarks in the fight by Negroes for equality. We now come to the more specific question, namely, the field of education. The question of the right of the state to practice segregation by race in certain educational facilities has only recently been tested in the courts. The cases of *Missouri ex rel. Gaines v. Canada* . . . and *Sipuel v. Board of Regents* . . . decided that Negroes were entitled to the same type of legal education that whites were given. It was further decided that the equal facilities must be furnished without delay or as was said in the *Sipuel* case, the state must provide for equality of education for Negroes "as soon as it does for applicants of any other group." But still we have not reached the exact question that is posed in the instant case.

We now come to the cases that, in my opinion, definitely and conclusively establish the doctrine that separation and segregation according to race is a violation of the Fourteenth Amendment. I, of course, refer to the cases of *Sweatt v. Painter* . . . and *McLaurin v. Oklahoma State Regents.* . . . These cases have been followed in a number of lower court decisions so that there is no longer any question as to the rights of Negroes to enjoy all the rights and facilities afforded by the law schools of the States of Virginia, Louisiana, Delaware, North Carolina and Kentucky. So there is no longer any basis for a state to claim the power to separate according to race in graduate schools, universities and colleges. . . .

. . . It has been said and repeated here in argument that the Supreme Court has refused to review the *Plessy* case in the *Sweatt, McLaurin* and other cases and this has been pointed to as proof that the Supreme Court retains and approves the validity of *Plessy*. It is astonishing that such an

---

[5]Labor: Steele v. Louisville & N. R. R. Co., 323 U.S. 192, 65 S.Ct. 226, 89 L.Ed. 173; Tunstall v. Brotherhood of Locomotive Firemen, 323 U.S. 210, 65 S.Ct. 235, 89 L.Ed. 187.

[6]Suffrage: Guinn v. U.S., 238 U.S. 347, 35 S.Ct. 926, 59 L.Ed. 1340; Nixon v. Herndon, 273 U.S. 536, 47 S.Ct. 446, 71 L.Ed. 759; Lane v. Wilson, 307 U.S. 268, 59 S.Ct. 872, 83 L.Ed. 1281; Smith v. Allwright, 321 U.S. 649, 64 S.Ct. 757, 88 L.Ed. 987; Elmore v. Rice, D.C., 72 F.Supp. 516; 4 Cir., 165 F.2d 387; certiorari denied, 333 U.S. 875, 68 S.Ct. 905, 92 L.Ed. 1151; Brown v. Baskin, D.C., 78 F.Supp. 933; Brown v. Baskin, D.C., 80 F.Supp. 1017; 4 Cir., 174 F.2d 391.

argument should be presented or used in this or any other court. . . . It was not considering railroad matters, had no arguments in regard to it, had no business or concern with railroad accommodations and should not have even been asked to refer to that case since it had no application or business in the consideration of an educational problem before the court. It seems to me that we have already spent too much time and wasted efforts in attempting to show any similarity between traveling in a railroad coach in the confines of a state and furnishing education to the future citizens of this country. . . .

In the instant case, the plaintiffs produced a large number of witnesses. It is significant that the defendants brought but two. These last two were not trained educators. One was an official of the Clarendon schools who said that the school system needed improvement and that the school officials were hopeful and expectant of obtaining money from State funds to improve all facilities. The other witness, significantly named Crow, has been recently employed by a commission just established which, it is proposed, will supervise educational facilities in the State and will handle monies if, as and when the same are received sometime in the future. Mr. Crow did not testify as an expert on education although he stated flatly that he believed in separation of the races and that he heard a number of other people say so, including some Negroes, but he was unable to mention any of their names. Mr. Crow explained what was likely and liable to happen under the 1951 State Educational Act to which frequent reference was made in argument on behalf of the defense.

It appears that the Governor of this state called upon the legislature to take action in regard to the dearth of educational facilities in South Carolina pointing out the low depth to which the state had sunk. As a result, an act of the legislature was adopted (this is part of the General Appropriations Act adopted at the recent session of the legislature and referred to as the 1951 School Act). This Act provides for the appointment of a commission which is to generally supervise educational facilities and imposes sales taxes in order to raise money for educational purposes and authorizes the issuance of bonds not to exceed the sum of $75,000,000, for the purpose of making grants to various counties and school districts to defray the cost of capital improvements in schools. . . . Nowhere is it specifically provided that there shall be equality of treatment as between whites and Negroes in the school system. It is openly and frankly admitted by all parties that the present facilities are hopelessly disproportional and no one knows how much money would be required to bring the colored school system up to a parity with the white school system. The estimates as to the cost merely of equalization of physical facilities run anywhere from forty to eighty million dollars. Thus, the position of the defendants is that the rights applied for by the plaintiffs are to be denied now because the State of South Carolina intends (as evidenced by a general appropriations bill enacted by the legislature and a speech made by its Governor) to issue bonds, impose taxes, raise money and to do something about the inadequate schools in the future. There is no guarantee or assurance as to when the money will be available. As yet, no bonds have been printed or sold. No money is in the treasury. No plans have been drawn for school buildings or order issued for materials. No allocation has been made to the Clarendon school district or any other school districts and not even application blanks have, as yet, been printed. But according to Mr. Crow, the Clarendon authorities have requested him to send them blanks for this purpose if, as and when they come into being. Can we seriously consider this a bona-fide attempt to provide equal facilities for our school children?

On the other hand, the plaintiffs brought many witnesses, some of them of national reputation in various educational fields. It is unnecessary for me to review or analyze their testimony. But they who had made studies of education and its effect upon children, starting with the lowest grades and studying them up through and into high school, unequivocally testified that aside from inequality in housing appliances and equipment, the mere fact of segregation, itself, had a deleterious and warping effect upon the minds of children. These witnesses testified as to their study and researches and their actual tests with children of varying ages and they showed that the humiliation and disgrace of being set aside and segregated as unfit to associate with others of different color had an evil and ineradicable effect upon the mental processes of our young which would remain with them and deform their view on life until and throughout their maturity. This applies to white as well as Negro children. These witnesses testified from actual study and tests in various parts of the country, including tests in the actual Clarendon School district under consideration. They showed beyond a doubt that the evils of segregation and color prejudice come from early training. And from their testimony as well as from common experience and knowledge and from our own reasoning, we must unavoidably come to the conclusion that racial prejudice is something that is acquired and that that acquiring is in early childhood. When do we get our first ideas of religion, nationality and the other basic ideologies? The vast number of individuals follow religious and political groups because of their childhood training. And it is difficult and nearly impossible to change and eradicate these early prejudices, however strong may be the appeal to reason. . . . If segregation is wrong then the place to stop it is in the first grade and not in graduate colleges.

From their testimony, it was clearly apparent, as it should be to any thoughtful person, irrespective of having such expert testimony, that segregation in education can never produce equality and that it is an evil that must be eradicated. This case presents the matter clearly for adjudication and I am of the opinion that all of the legal guideposts, expert testimony, common sense and reason point unerringly to the conclusion that the system of segregation in education adopted and practiced in the State of South Carolina must go and must go now.

Segregation is per se inequality. . . .

THE SUPREME COURT ROUNDS: THE MAKING OF *BROWN I* AND *BROWN II*

Round One: Setting the Stage

Appellants' Brief
1952

When the five consolidated cases reached the Supreme Court in late 1952, the contending positions had been well honed. The appellant's (plaintiff's appeal) and appellee's (defendant's appeal) briefs formally articulated the original oral arguments of the Brown *case, the appellants arguing that Jim Crow schools were a violation of the schoolchildren's right to equality before the law and should forthwith be replaced with integrated schools and the appellees arguing that Jim Crow schools were consistent with law and custom. The appellants wanted* Plessy *overturned; the appellees argued for confirming* Plessy *and reforming segregated schools by making the black schools equal to the white ones. The excerpts are from the Kansas case.*

The appendix to the appellants' brief shows the pioneering and effective use of social-scientific evidence. The issue of sociological jurisprudence caused concern in many quarters. What are the thesis and supporting argu-

Philip B. Kurland and Gerhard Casper, eds., *Landmark Briefs and Arguments of the Supreme Court of the United States: Constitutional Law*, vol. 49 (Arlington, Va.: University Publications of America, 1975), 31–32, 34–39.

ments in the appellants' appendix? What do you see as its strengths and weaknesses? Do you find it persuasive overall? Why or why not?

Comparing and contrasting the appellees' brief and the appellants' brief, which positions do you find more or less cogent? Why? How do these arguments compare and contrast with those in previous cases?

The Fourteenth Amendment precludes a state from imposing distinctions or classifications based upon race and color alone. The State of Kansas has no power thereunder to use race as a factor in affording educational opportunities to its citizens.

Racial segregation in public schools reduces the benefits of public education to one group solely on the basis of race and color and is a constitutionally proscribed distinction. Even assuming that the segregated schools attended by appellants are not inferior to other elementary schools in Topeka with respect to physical facilities, instruction and courses of study, unconstitutional inequality inheres in the retardation of intellectual development and distortion of personality which Negro children suffer as a result of enforced isolation in school from the general public school population. Such injury and inequality are established as facts on this appeal by the uncontested findings of the District Court.

The District Court reasoned that it could not rectify the inequality that it had found because of this Court's decisions in *Plessy* v. *Ferguson, . . .* and *Gong Lum* v. *Rice. . . .* This Court has already decided that the *Plessy* case is not in point. Reliance upon *Gong Lum* v. *Rice* is mistaken since the basic assumption of that case is the existence of equality while no such assumption can be made here in the face of the established facts. Moreover, more recent decisions of this Court, most notably *Sweatt* v. *Painter . . .* and *McLaurin* v. *Board of Regents . . .* clearly show that such hurtful consequences of segregated schools as appear here constitute a denial of equal educational opportunities in violation of the Fourteenth Amendment. Therefore, the court below erred in denying the relief prayed by appellants. . . .

The court below, having found that appellants were denied equal educational opportunities by virtue of the segregated school system, erred in denying the relief prayed.

The court below made the following finding of fact:

> Segregation of white and colored children in public schools has a detrimental effect upon the colored children. The impact is greater when

it has the sanction of the law; for the policy of separating the races is usually interpreted as denoting the inferiority of the negro group. A sense of inferiority affects the motivation of a child to learn. Segregation with the sanction of law, therefore, has a tendency to retard the educational and mental development of negro children and to deprive them of some of the benefits they would receive in a racially integrated school system.

This finding is based upon uncontradicted testimony that conclusively demonstrates that racial segregation injures infant appellants in denying them the opportunity available to all other racial groups to learn to live, work and cooperate with children representative of approximately 90% of the population of the society in which they live . . . ; to develop citizenship skills; and to adjust themselves personally and socially in a setting comprising a cross-section of the dominant population. . . . The testimony further developed the fact that the enforcement of segregation under law denies to the Negro status, power and privilege . . . ; interferes with his motivation for learning . . . ; and instills in him a feeling of inferiority . . . resulting in a personal insecurity, confusion and frustration that condemns him to an ineffective role as a citizen and member of society. . . . Moreover, it was demonstrated that racial segregation is supported by the myth of the Negro's inferiority . . . , and where, as here, the state enforces segregation, the community at large is supported in or converted to the belief that this myth has substance in fact. . . . It was testified that because of the peculiar educational system in Kansas that requires segregation only in the lower grades, there is an additional injury in that segregation at an early age is greater in its impact and more permanent in its effects . . . even though there is a change to integrated schools at the upper levels.

That these conclusions are the consensus of social scientists is evidenced by the appendix filed herewith. Indeed, the findings of the court that segregation constitutes discrimination are supported on the face of the statute itself. . . .

Under the Fourteenth Amendment equality of educational opportunities necessitates an evaluation of all factors affecting the educational process. . . . Applying this yardstick, any restrictions or distinction based upon race or color that places the Negro at a disadvantage in relation to other racial groups in his pursuit of educational opportunities is violative of the equal protection clause.

In the instant case, the court found as a fact that appellants were placed at such a disadvantage and were denied educational opportunities equal to those available to white students. It necessarily follows, therefore, that the court should have concluded as a matter of law that appellants were deprived of their right to equal educational opportunities in violation of the equal protection clause of the Fourteenth Amendment. Under the mistaken notion that *Plessy* v. *Ferguson* and *Gong Lum* v. *Rice* were controlling with respect to the validity of racial distinctions in elementary education, the trial court refused to conclude that appellants were here denied equal educational opportunities in violation of their constitutional rights. Thus, notwithstanding that it had found inequality in educational opportunity as a fact, the court concluded as a matter of law that such inequality did not constitute a denial of constitutional rights, saying:

> *Plessy* v. *Ferguson* . . . and *Gong Lum* v. *Rice* . . . uphold the constitutionality of a legally segregated school system in the lower grades and no denial of due process results from the maintenance of such a segregated system of schools absent discrimination in the maintenance of such a segregated system of schools. We conclude that the above-cited cases have not been overruled by the later case of *McLaurin* v. *Oklahoma* . . . and *Sweatt* v. *Painter.* . . .

Plessy v. *Ferguson* is not applicable. Whatever doubts may once have existed in this respect were removed by this Court in *Sweatt* v. *Painter.* . . .

Gong Lum v. *Rice* is irrelevant to the issues in this case. There, a child of Chinese parentage was denied admission to a school maintained exclusively for white children and was ordered to attend a school for Negro children. The power of the state to make racial distinctions in its school system was not in issue. Petitioner contended that she had a constitutional right to go to school with white children, and that in being compelled to attend school with Negroes, the state had deprived her of the equal protection of the laws.

Further, there was no showing that her educational opportunities had been diminished as a result of the state's compulsion, and it was assumed by the Court that equality in fact existed. There the petitioner was not inveighing against the system, but that its application resulted in her classification as a Negro rather than as a white person, and indeed by so much conceded the propriety of the system itself. Were this not true, this Court would not have found basis for holding that the issue raised was one "which has been many times decided to be within the constitutional power of the state" and, therefore, did not "call for very full argument and consideration."

In short, she raised no issue with respect to the state's power to enforce racial classifications, as do appellants here. Rather, her objection went only to her treatment under the classification. This case, therefore, cannot be pointed to as a controlling precedent covering the instant case in which the constitutionality of the system itself is the basis for attack and in which it is shown the inequality in fact exists.

In any event the assumptions in the *Gong Lum* case have since been rejected by this Court. In the *Gong Lum* case, without "full argument and consideration," the Court assumed the state had power to make racial distinctions in its public schools without violating the equal protection clause of the Fourteenth Amendment and assumed the state and lower federal court cases cited in support of this assumed state power had been correctly decided. Language in *Plessy* v. *Ferguson* was cited in support of these assumptions. These assumptions upon full argument and consideration were rejected in the *McLaurin* and *Sweatt* cases in relation to racial distinctions in state graduate and professional education. And, according to those cases, *Plessy* v. *Ferguson,* is not controlling for the purpose of determining the state's power to enforce racial segregation in public schools.

Thus, the very basis of the decision in the *Gong Lum* case has been destroyed. We submit, therefore, that this Court has considered the basic issue involved here only in those cases dealing with racial distinctions in education at the graduate and professional levels. . . .

In the *McLaurin* and *Sweatt* cases, this Court measured the effect of racial restrictions upon the educational development of the individual affected, and took into account the community's actual evaluation of the schools involved. In the instant case, the court below found as a fact that racial segregation in elementary education denoted the inferiority of Negro children and retarded their educational and mental development. Thus the same factors which led to the result reached in the *McLaurin* and *Sweatt* cases are present. Their underlying principles, based upon sound analyses, control the instant case.

Conclusion

In light of the foregoing, we respectfully submit that appellants have been denied their rights to equal educational opportunities within the meaning of the Fourteenth Amendment and that the judgment of the court below should be reversed.

The Effects of Segregation and the Consequences of Desegregation: A Social Science Statement

Appendix to Appellants' Brief

I

The problem of the segregation of racial and ethnic groups constitutes one of the major problems facing the American people today. It seems desirable, therefore, to summarize the contributions which contemporary social science can make toward its resolution. There are, of course, moral and legal issues involved with respect to which the signers of the present statement cannot speak with any special authority and which must be taken into account in the solution of the problem. There are, however, also factual issues involved with respect to which certain conclusions seem to be justified on the basis of the available scientific evidence. It is with these issues only that this paper is concerned. Some of the issues have to do with the consequences of segregation, some with the problems of changing from segregated to unsegregated practices. These two groups of issues will be dealt with in separate sections below. It is necessary, first, however, to define and delimit the problem to be discussed.

Definitions

For purposes of the present statement, *segregation* refers to that restriction of opportunities for different types of associations between the members of one racial, religious, national or geographic origin, or linguistic group and those of other groups, which results from or is supported by the action of any official body or agency representing some branch of government. . . .

Where the action takes place in a social milieu in which the groups involved do not enjoy equal social status, the group that is of lesser social status will be referred to as the *segregated* group.

In dealing with the question of the effects of segregation, it must be

Philip B. Kurland and Gerhard Casper, eds., *Landmark Briefs and Arguments of the Supreme Court of the United States: Constitutional Law,* vol. 49 (Arlington, Va.: University Publications of America, 1975), 43–60.

recognized that these effects do not take place in a vacuum, but in a social context. The segregation of Negroes and of other groups in the United States takes place in a social milieu in which "race" prejudice and discrimination exist. . . . The imbeddedness of segregation in such a context makes it difficult to disentangle the effects of segregation *per se* from the effects of the context. . . . We shall, however, return to this problem after consideration of the observable effects of the total social complex in which segregation is a major component.

II

At the recent Mid-century White House Conference on Children and Youth, a fact-finding report on the effects of prejudice, discrimination and segregation on the personality development of children was prepared as a basis for some of the deliberations.[1] This report brought together the available social science and psychological studies which were related to the problem of how racial and religious prejudices influenced the development of a healthy personality. It highlighted the fact that segregation, prejudices and discriminations, and their social concomitants potentially damage the personality of all children—the children of the majority group in a somewhat different way than the more obviously damaged children of the minority group.

The report indicates that as minority group children learn the inferior status to which they are assigned—as they observe the fact that they are almost always segregated and kept apart from others who are treated with more respect by the society as a whole—they often react with feelings of inferiority and a sense of personal humiliation. Many of them become confused about their own personal worth. On the one hand, like all other human beings they require a sense of personal dignity; on the other hand, almost nowhere in the larger society do they find their own dignity as human beings respected by others. Under these conditions, the minority group child is thrown into a conflict with regard to his feelings about himself and his group. He wonders whether his group and he himself are worthy of no more respect than they receive. This conflict and confusion leads to self-hatred and rejection of his own group. . . .

Some children, usually of the lower socio-economic classes, may react by overt aggressions and hostility directed toward their own group or

[1]Clark, K.B., *Effect of Prejudice and Discrimination on Personality Development,* Fact Finding Report Mid-century White House Conference on Children and Youth, Children's Bureau, Federal Security Agency, 1950 (mimeographed).

members of the dominant group.[2] Anti-social and delinquent behavior may often be interpreted as reactions to these racial frustrations. These reactions are self-destructive in that the larger society not only punishes those who commit them, but often interprets such aggressive and anti-social behavior as justification for continuing prejudice and segregation.

Middle class and upper class minority group children are likely to react to their racial frustrations and conflicts by withdrawal and submissive behavior. Or, they may react with compensatory and rigid conformity to the prevailing middle class values and standards and an aggressive determination to succeed in these terms in spite of the handicap of their minority status.

The report indicates that minority group children of all social and economic classes often react with a generally defeatist attitude and a lowering of personal ambitions. This, for example, is reflected in a lowering of pupil morale and a depression of the educational aspiration level among minority group children in segregated schools. In producing such effects, segregated schools impair the ability of the child to profit from the educational opportunities provided him.

Many minority group children of all classes also tend to be hypersensitive and anxious about their relations with the larger society. They tend to see hostility and rejection even in those areas where these might not actually exist. The report concludes that while the range of individual differences among members of a rejected minority group is as wide as among other peoples, the evidence suggests that all of these children are unnecessarily encumbered in some ways by segregation and its concomitants.

With reference to the impact of segregation and its concomitants on children of the majority group, the report indicates that the effects are somewhat more obscure. Those children who learn the prejudices of our society are also being taught to gain personal status in an unrealistic and non-adaptive way. When comparing themselves to members of the minority group, they are not required to evaluate themselves in terms of the more basic standards of actual personal ability and achievement. The culture permits and, at times, encourages them to direct their feelings of hostility and aggression against whole groups of people the members of

[2]Brenman, M., The Relationship Between Minority Group Identification in a Group of Urban Middle Class Negro Girls, *J. Soc. Psychol.*, 1940, 11, 171–197; Brenman, M., Minority Group Membership and Religious, Psychosexual and Social Patterns in a Group of Middle-Class Negro Girls, *J. Soc. Psychol.*, 1940, 12, 179–196; Brenman, M., Urban Lower-Class Negro Girls, *Psychiatry*, 1943, 6, 307–324; Davis, A., The Socialization of the American Negro Child and Adolescent, *J. Negro Educ.*, 1939, 8, 264–275.

which are perceived as weaker than themselves. They often develop patterns of guilt feelings, rationalizations and other mechanisms which they must use in an attempt to protect themselves from recognizing the essential injustice of their unrealistic fears and hatreds of minority groups.[3]

The report indicates further that confusion, conflict, moral cynicism, and disrespect for authority may arise in majority group children as a consequence of being taught the moral, religious and democratic principles of the brotherhood of man and the importance of justice and fair play by the same persons and institutions who, in their support of racial segregation and related practices, seem to be acting in a prejudiced and discriminatory manner. Some individuals may attempt to resolve this conflict by intensifying their hostility toward the minority group. Others may react by guilt feelings which are not necessarily reflected in more humane attitudes toward the minority group. Still others react by developing an unwholesome, rigid, and uncritical idealization of all authority figures — their parents, strong political and economic leaders. As described in *The Authoritarian Personality*,[4] they despise the weak, while they obsequiously and unquestioningly conform to the demands of the strong whom they also, paradoxically, subconsciously hate. . . .

Conclusions similar to those reached by the Mid-century White House Conference Report have been stated by other social scientists who have concerned themselves with this problem. The following are some examples of these conclusions:

Segregation imposes upon individuals a distorted sense of social reality.[5]

Segregation leads to a blockage in the communications and interaction between the two groups. Such blockages tend to increase mutual suspicion, distrust and hostility.[6]

Segregation not only perpetuates rigid stereotypes and reinforces negative attitudes toward members of the other group, but also leads to the development of a social climate within which violent outbreaks of racial tensions are likely to occur.[7]

We return now to the question, deferred earlier, of what it is about the

---

[3]Adorno, T.W.; Frenkel-Brunswik, E.; Levinson, D.J.; Sanford, R.N., *The Authoritarian Personality*, 1951.

[4]Adorno, T.W.; Frenkel-Brunswik, E.; Levinson, D.J.; Sanford, R.N., *The Authoritarian Personality*, 1951.

[5]Reid, Ira, What Segregated Areas Mean; Brameld, T., Educational Cost, *Discrimination and National Welfare*, Ed. by MacIver, R.M., 1949.

[6]Frazier, E., *The Negro in the United States*, 1949; Krech, D. & Crutchfield, R. S., *Theory and Problems of Social Psychology*, 1948; Newcomb, T., *Social Psychology*, 1950.

[7]Lee, A. McClung and Humphrey, N.D., *Race Riot*, 1943.

total society complex of which segregation is one feature that produces the effects described above—or, more precisely, to the question of whether we can justifiably conclude that, as only one feature of a complex social setting, segregation is in fact a significantly contributing factor to these effects.

To answer this question, it is necessary to bring to bear the general fund of psychological and sociological knowledge concerning the role of various environmental influences in producing feelings of inferiority, confusions in personal roles, various types of basic personality structures and the various forms of personal and social disorganization.

On the basis of this general fund of knowledge, it seems likely that feelings of inferiority and doubts about personal worth are attributable to living in an underprivileged environment only insofar as the latter is itself perceived as an indicator of low social status and as a symbol of inferiority. In other words, one of the important determinants in producing such feelings is the awareness of social status difference. While there are many other factors that serve as reminders of the differences in social status, there can be little doubt that the fact of enforced segregation is a major factor.[8]

This seems to be true for the following reasons among others: (1) because enforced segregation results from the decision of the majority group without the consent of the segregated and is commonly so perceived; and (2) because historically segregation patterns in the United States were developed on the assumption of the inferiority of the segregated.

In addition, enforced segregation gives official recognition and sanction to these other factors of the social complex, and thereby enhances the effects of the latter in creating the awareness of social status differences and feelings of inferiority.[9] The child who, for example, is compelled to attend a segregated school may be able to cope with ordinary expressions of prejudice by regarding the prejudiced person as evil or misguided; but he cannot readily cope with symbols of authority, the full force of the authority of the State—the school or the school board, in this instance—in the same manner. Given both the ordinary expression of prejudice and the school's policy of segregation, the former takes on greater force and seemingly becomes an official expression of the latter.

Not all of the psychological traits which are commonly observed in the social complex under discussion can be related so directly to the aware-

---

[8]Frazier, E., *The Negro in the United States,* 1949; Myrdal, G., *An American Dilemma,* 1944.
    [9]Reid, Ira, What Segregated Areas Mean, *Discrimination and National Welfare,* Ed. by MacIver, R.M., 1949.

ness of status differences—which in turn is, as we have already noted, materially contributed to by the practices of segregation. Thus, the low level of aspiration and defeatism so commonly observed in segregated groups is undoubtedly related to the level of self-evaluation; but it is also, in some measure, related among other things to one's expectations with regard to opportunities for achievement and, having achieved, to the opportunities for making use of these achievements. Similarly, the hypersensitivity and anxiety displayed by many minority group children about their relations with the larger society probably reflects their awareness of status differences; but it may also be influenced by the relative absence of opportunities for equal status contact which would provide correctives for prevailing unrealistic stereotypes.

The preceding view is consistent with the opinion stated by a large majority (90%) of social scientists who replied to a questionaire concerning the probable effects of enforced segregation under conditions of equal facilities. This opinion was that, regardless of the facilities which are provided, enforced segregation is psychologically detrimental to the members of the segregated group. . . . [10]

It may be noted that many of these social scientists supported their opinions on the effects of segregation on both majority and minority groups by reference to one or another or to several of the following four lines of published and unpublished evidence.[11] First, studies of children throw light on the relative priority of the awareness of status differentials and related factors as compared to the awareness of differences in facilities. On this basis, it is possible to infer some of the consequences of segregation as distinct from the influence of inequalities of facilities. Second, clinical studies and depth interviews throw light on the genetic sources and causal sequences of various patterns of psychological reaction; and, again, certain inferences are possible with respect to the effects of segregation *per se* in situations where one finds both segregation and unequal facilities.

III

Segregation is at present a social reality. Questions may be raised, therefore, as to what are the likely consequences of desegregation.

One such question asks whether the inclusion of an intellectually inferior group may jeopardize the education of the more intelligent group

---

[10]Deutscher, M. and Chein, I., The Psychological Effects of Enforced Segregation: A Survey of Social Science Opinion, *J. Psychol.,* 1948, 26, 259–287.

[11]Chein, I., What Are the Psychological Effects of Segregation Under Conditions of Equal Facilities?, *International J. Opinion and Attitude Res.,* 1949, 2, 229–234.

by lowering educational standards or damage the less intelligent group by placing it in a situation where it is at a marked competitive disadvantage. Behind this question is the assumption, which is examined below, that the presently segregated groups actually are inferior intellectually.

The available scientific evidence indicates that much, perhaps all, of the observable differences among various racial and national groups may be adequately explained in terms of environmental differences.[12] It has been found, for instance, that the differences between the average intelligence test scores of Negro and white children decrease, and the overlap of the distributions increases, proportionately to the number of years that the Negro children have lived in the North.[13] Related studies have shown that this change cannot be explained by the hypothesis of selective migration.[14] It seems clear, therefore, that fears based on the assumption of innate racial differences in intelligence are not well founded.

It may also be noted in passing that the argument regarding the intellectual inferiority of one group as compared to another is, as applied to schools, essentially an argument for homogeneous groupings of children by intelligence rather than by race. . . . [M]any educators have come to doubt the wisdom of class groupings made homogeneous solely on the basis of intelligence.[15] Those who are opposed to such homogeneous grouping believe that this type of segregation, too, appears to create generalized feelings of inferiority in the child who attends a below average class, leads to undesirable emotional consequences in the education of the gifted child, and reduces learning opportunities which result from the interaction of individuals with varied gifts.

A second problem that comes up in an evaluation of the possible consequences of desegregation involves the question of whether segregation prevents or stimulates interracial tension and conflict and the corollary question of whether desegregation has one or the other effect.

The most direct evidence available on this problem comes from observations and systematic study of instances in which desegregation has

---

[12]Klineberg, O. *Characteristics of American Negro,* 1945; Klineberg, O., *Race Differences,* 1936.
[13]Klineberg, O. *Negro Intelligence and Selective Migration,* 1935.
[14]Klineberg, O. *Negro Intelligence and Selective Migration,* 1935.
[15]Brooks, J.J., Interage Grouping on Trial-Continuous Learning, *Bulletin #87, Association for Childhood Education,* 1951; Lane, R.H., Teacher in Modern Elementary School, 1941; Educational Policies Commission of the National Education Association and the American Association of School Administration Report in *Education For All Americans,* published by the N.E.A. 1948.

occurred. Comprehensive reviews of such instances[16] clearly establish the fact that desegregation has been carried out successfully in a variety of situations although outbreaks of violence had been commonly predicted. Extensive desegregation has taken place without major incidents in the armed services in both Northern and Southern installations and involving officers and enlisted men from all parts of the country, including the South.[17] Similar changes have been noted in housing[18] and in industry.[19] During the last war, many factories both in the North and South hired Negroes on a non-segregated, non-discriminatory basis. While a few strikes occurred, refusal by management and unions to yield quelled all strikes within a few days.[20]

Relevant to this general problem is a comprehensive study of urban race riots which found that race riots occurred in segregated neighborhoods, whereas there was no violence in sections of the city where the two races lived, worked and attended school together.[21]

Under certain circumstances desegregation not only proceeds without major difficulties, but has been observed to lead to the emergence of more favorable attitudes and friendlier relations between races. . . .

Much depends, however, on the circumstances under which members of previously segregated groups first come in contact with others in unsegregated situations. Available evidence suggests, first, that there is less likelihood of unfriendly relations when the change is simultaneously

[16]Delano, W., Grade School Segregation: The Latest Attack on Racial Discrimination, *Yale Law Journal*, 1952, 61, 5, 730–744; Rose, A., The Influence of Legislation on Prejudice; Chapter 53 in *Race Prejudice and Discrimination*, Ed. by Rose, A., 1951; Rose, A., *Studies in the Reduction of Prejudice*, Amer. Council on Race Relations, 1948.

[17]Kenworthy, E.W., The Case Against Army Segregation, *Annals of the American Academy of Political and Social Science*, 1951, 275, 27–33; Nelson, Lt. D.D., *The Integration of the Negro in the U.S. Navy*, 1951; Opinions About Negro Infantry Platoons in White Companies in Several Divisions, *Information and Education Division, U.S. War Department, Report No. B-157*, 1945.

[18]Conover, R.D., *Race Relations at Codornices Village, Berkeley-Albany, California: A Report of the Attempt to Break Down the Segregated Pattern on a Directly Managed Housing Project*, Housing and Home Finance Agency, Public Housing Administration, Region I, December 1947 (mimeographed); Deutch, M. and Collins, M.E., *Interracial Housing, A Psychological Study of a Social Experiment*, 1951; Rutledge, E., *Integration of Racial Minorities in Public Housing Projects: A Guide for Local Housing Authorities on How to Do It*, Public Housing Administration, New York Field Office (mimeographed).

[19]Minard, R.D., The Pattern of Race Relationships in the Pocahontas Coal Field, *J. Social Issues*, 1952, 8, 29–44; Southall, S.E., *Industry's Unfinished Business*, 1951; Weaver, G. L-P, *Negro Labor, A National Problem*, 1941.

[20]Southall, S.E., *Industry's Unfinished Business*, 1951; Weaver, G. L-P, *Negro Labor, A National Problem*, 1941.

[21]Lee, A. McClung and Humphrey, N.D., *Race Riot*, 1943; Lee, A. McClung, Race Riots Aren't Necessary, *Public Affairs Pamphlet*, 1945.

introduced into all units of a social institution to which it is applicable—
e.g., all of the schools in a school system or all of the shops in a given fac-
tory.[22] When factories introduced Negroes in only some shops but not in
others the prejudiced workers tended to classify the desegregated shops
as inferior, "Negro work." Such objections were not raised when complete
integration was introduced.

The available evidence also suggests the importance of consistent and
firm enforcement of the new policy by those in authority.[23] It indicates
also the importance of such factors as: the absence of competition for a
limited number of facilities or benefits;[24] the possibility of contacts which
permit individuals to learn about one another as individuals;[25] and the possibility of equivalence of positions and functions among all of the participants within the unsegregated situation.[26] These conditions can generally be satisfied in a number of situations, as in the armed services, public
housing developments, and public schools.

IV

The problem with which we have here attempted to deal is admittedly on
the frontiers of scientific knowledge. Inevitably, there must be some differences of opinion among us concerning the conclusiveness of certain
items of evidence, and concerning the particular choice of words and
placement of emphasis in the preceding statement. We are nonetheless
in agreement that this statement is substantially correct and justified by

---

[22]Minard, R.D., The Pattern of Race Relationships in the Pocahontas Coal Field, *J. Social Issues*, 1952, 8, 29–44; Rutledge, E., *Integration of Racial Minorities in Public Housing Projects: A Guide for Local Housing Authorities on How To Do It*, Public Housing Administration, New York Field Office (mimeographed).

[23]Deutsch, M. and Collins, M.E., *Interracial Housing, A Psychological Study of a Social Experiment*, 1951; Feldman, H., The Technique of Introducing Negroes Into the Plant, *Personnel*, 1942, 19, 461–466; Rutledge, E., *Integration of Racial Minorities in Public Housing Projects: A Guide for Local Housing Authorities on How To Do It*, Public Housing Administration, New York Field Office (mimeographed); Southall, S.E., *Industry's Unfinished Business*, 1951; Watson, G., *Action for Unity*, 1947.

[24]Lee, A. McClung and Humphrey, N.D., *Race Riot*, 1943; Williams, R., Jr., *The Reduction of Intergroup Tensions*, Social Science Research Council, New York, 1947; Windner, A.E., *White Attitudes Towards Negro-White Interaction in An Area of Changing Racial Composition*. Paper Delivered at the Sixtieth Annual Meeting of the American Psychological Association, Washington, September 1952.

[25]Wilner, D.M.; Walkley, R.P.; and Cook, S.W., Intergroup Contact and Ethnic Attitudes in Public Housing Projects, *J. Social Issues*, 1952, 8, 45–69.

[26]Allport, G.W., and Kramer, B., Some Roots of Prejudice, *J. Psychol.*, 1946, 22, 9–39; Watson, J., Some Social and Psychological Situations Related to Change in Attitude, *Human Relations*, 1950, 3, 1.

the evidence, and the differences among us, if any, are of a relatively minor order and would not materially influence the preceding conclusions.

Appellees' Brief
1952

The issue presented by this case is whether the Fourteenth Amendment to the Constitution of the United States is violated by a statute which permits boards of education in designated cities to maintain separate elementary school facilities for the education of white and colored children.

At the outset, counsel for the appellees desire to state that by appearing herein they do not propose to advocate the policy of segregation of any racial group within the public school system. We contend only that policy determinations are matters within the exclusive province of the legislature. We do not express an opinion as to whether the practice of having separate schools of equal facility for the white and colored races is economically expedient or sociologically desirable, or whether it is consistent with sound ethical or religious theory. We do not understand that these extra-legal questions are now before the Court. The only proposition that we desire to urge is that the Kansas statute which permits racial segregation in elementary public schools in certain cities of the state does not violate the Fourteenth Amendment to the Constitution of the United States as that amendment has been interpreted and applied by this Court.

. . . The Court below found facilities provided for Negro children in the city of Topeka to be substantially equal to those furnished to white children. The appellants, in their specifications of error and in their brief, do not object to that finding. Under those circumstances and under authority of the decisions of the Supreme Court of the United States, the inferior federal courts, and the courts of last resort in numerous state jurisdictions, and particularly the decisions of the Kansas Supreme Court, the appellants herein are not denied equal protection of the laws by virtue of their being required to attend schools separate from those which white children are required to attend.

The decision of the court below should be affirmed. . . .

Philip B. Kurland and Gerhard Casper, eds., *Landmark Briefs and Arguments of the Supreme Court of the United States: Constitutional Law,* vol. 49 (Arlington, Va.: University Publications of America, 1975), 70–71, 75–76, 98–100, 102–4, 106–9.

The Prospect

At the outset we suggested that the Kansas statute is permissive and that any Board of Education included in the statute may adopt a policy consistent with local conditions and local attitudes. We believe it is significant that under this statute by a process of evolution the people in Kansas communities are arriving at their own solutions to this problem. Under the statute fourteen cities are authorized to maintain separate schools for colored students. The files of the State Superintendent of Public Instruction indicate that at the present time, only nine cities exercise the power conferred by statute. . . .

This account of events not in the record is related to illustrate the wisdom which underlies the Kansas statute. Only those cities where local conditions produce special problems making segregation desirable need adopt the expedient of segregation. In the orderly progress of the community, these special problems are either solved or vanish, and when the need for segregation disappears, its practice may be discontinued. This was the method provided by the legislature of the State of Kansas to achieve the goal of an integrated school system where segregation is not needed. We respectfully suggest to the court that this evolutionary process permitting an autonomous solution in the community is consistent with the purpose and intent of the Fourteenth Amendment. . . .

The District Court's Finding of Fact No. VIII Is Insufficient to Establish Appellants' Right to Injunctive Relief and to Require Reversal of the Judgment Below.

(A) COUNSEL FOR APPELLANTS HAVE OVERSTATED THEIR CASE. . . .

With all respect due to able counsel for appellants we believe that in their zeal for their cause, they have overstated their case. The only existing Finding of Fact which is relied upon by appellants and the only one quoted in their brief is the District Court's Finding of Fact No. VIII, which we quote accurately:

> Segregation of white and colored children in public schools has a detrimental effect upon the colored children. The impact is greater when it has the sanction of the law; for the policy of separating the races is usually interpreted as denoting the inferiority of the Negro group. A sense of inferiority affects the motivation of a child to learn. Segrega-

tion with the sanction of law, therefore, has a tendency to retain the educational and mental development of Negro children and to deprive them of some of the benefits they would receive in a racial integrated school system.

We call attention to the fact that the foregoing Finding is couched only in broad and general language; it makes no specific or particular reference to any of the appellants, nor to the grade schools in Topeka, nor to racial groups other than Negroes, nor to inequality of educational opportunities between Negroes and other racial groups. The substance of the finding can be summarized in the following statement: "Generally speaking, segregation is detrimental to colored children, and deprives them of some benefits they would receive in a racially integrated school system."

The finding of Fact No. VIII cannot be stretched, as counsel for appellants apparently would like to stretch it, into a finding that the appellants in this case have "suffered serious harm in being required to attend segregated elementary schools in Topeka" and that "appellants were placed at such a disadvantage (in relation to other racial groups in [their] pursuit of educational opportunities) and were denied educational opportunities equal to those available to white students."

(B) ELEMENTS NECESSARY TO ENTITLE APPELLANTS TO INJUNCTIVE RELIEF AND TO A REVERSAL OF THE JUDGMENT IN THIS CASE.

To establish appellants' right to injunctive relief and to reversal of the judgment in this case, the Findings of Fact No. VIII would have to show:

(1) That the appellants have actually suffered personal harm as the result of attending segregated schools in Topeka; and,

(2) Either that appellants are being deprived of benefits which other students in the Topeka school system enjoy, or that appellants are being subjected to detriments to which other students in the Topeka school system are not being subjected, by reason of maintenance of a segregated school system.

The mere showing that appellants may be members of a class which is being discriminated against by reason of a statute is not sufficient to entitle them to injunctive relief, unless appellants can also show that they personally are suffering harm. The Fourteenth Amendment protects only personal and individual rights.

The mere showing that appellants can show that they are being deprived of benefits they would receive under a different system of schools is not sufficient to show that they are being deprived of equal protection of the law unless appellants can also show that under the existing segregated school system there are others who are not deprived of such benefits.

And finally, the mere showing that segregation is detrimental to appellants is not sufficient to show that they are being deprived of equal protection of the laws, unless they also show that segregation is not similarly detrimental to others in the Topeka school system. . . .

(c) FINDING OF FACT NO. VIII FAILS TO DISCLOSE THAT ANY OF THE APPELLANTS HAVE BEEN ACTUALLY AND PERSONALLY HARMED BY SEGREGATION IN THE TOPEKA SCHOOLS.

Finding of Fact No. VIII makes no specific reference to the individual appellants. It expresses only in broad generalities the effect of segregation in the public schools upon colored children as a class. There is no specific finding that segregation has had a personal detrimental effect upon any of the appellants. There is no specific finding that any of the appellants personally has interpreted segregation as denoting inferiority of the Negro group, or that the motivation to learn of any of the appellants has been affected by a sense of inferiority. There is no finding that the educational and mental development of any of the appellants has actually been retained or retarded by reason of segregation in the Topeka schools. . . .

(d) FINDING OF FACT NO. VIII FAILS TO DISCLOSE THAT APPELLANTS ARE BEING DEPRIVED OF EQUAL PROTECTION OF THE LAWS, OR THAT THEY ARE BEING DISCRIMINATED AGAINST BY SEGREGATION IN THE TOPEKA SCHOOLS.

Denial of equal protection of the laws, or discrimination, logically and necessarily involves at least two persons who are being treated differently. Denial of equal protection must mean denial of protection or opportunity equal to that afforded to someone else. There can be no such thing as "unilateral discrimination."

Since the Finding of Fact No. VIII is limited solely to a statement of the effect of segregation on colored children as a group, and nowhere mentions the effect of segregation upon any other race or group, it cannot reasonably or logically show discrimination or a denial of equal protection of the laws.

Nowhere in the finding has the court disclosed any facts upon which it can be claimed to show discrimination in favor of white children over colored in segregated schools.

It is idle on this appeal to speculate upon what the trial court might have found had it been requested to make additional findings. No request for additional findings was made in the trial court. We therefore refrain from speculating as to whether the court would also have found that segregation was detrimental to white children and impaired their educational and mental development.

(E) THE DISTRICT COURT DID NOT INTEND NOR CONSIDER ITS FINDING OF FACT NO. VIII TO BE A FINDING OF DISCRIMINATION AGAINST APPELLANTS.

The last sentence in Finding of Fact No. VIII summarizes the entire finding. We quote:

> Segregation with the sanction of law, therefore, has a tendency to retain the educational and mental development of Negro children and to deprive them of some of the benefits they would receive in a racially integrated school system.

We believe the court intended the finding to mean simply that colored children would be better off in integrated schools than they are in segregated schools. Conceding that that is the meaning of the finding, it does not amount to a finding of actual discrimination against colored children and in favor of white children upon the facts in this case. White children are not permitted to attend integrated schools in Topeka. The mere fact, if it be a fact, that the Topeka school system could be improved so far as education of colored children is concerned, does not prove discrimination against them.

In the opinion of the District Court . . . , no mention is made of Finding of Fact No. VIII. It is clear the District Court did not consider or intend to attach to that finding the same significance which appellants seek to place upon it.

We do not question that if the Finding of Fact No. VIII means everything appellants claim it means, they would be entitled to an injunction and reversal of the judgment, if this court should overrule the "separate but equal doctrine." However, it is clear that the District Court did not intend or consider the finding to mean all the things appellants claim for it. As stated in the Decree of the District Court:

> The Court has heretofore filed its Findings of Fact and Conclusions of Law together with an opinion and has held as a matter of law that the plaintiffs have failed to prove they are entitled to the relief demanded.

**Round Two: Reargument on
Original Intent and Possible Relief**

The Supreme Court's Order:
The Questions
1953

*In this round, the case focused on the issue of the original intent of the leg-
islative framers of the Fourteenth Amendment: Did they intend through this
amendment to sustain or abolish segregated schools? Original intent serves
as an important guide for those concerned with interpreting a point of law
within the narrow, often conservative, confines of its original meaning.
This is not always as straightforward as it might appear because legislation
often results from complicated and multisided debates that reveal contrasting points of view and compromise results. Even relative consensus, as in
this instance, can often be interpreted quite differently, especially after the
fact. Nevertheless, the notion of original intent frames judicial tradition
through its insistence on the original historical moment as the most reliable
guide to constitutionality.*

*This point of view contrasts nicely with a contingent view of legal deci-
sion making that emphasizes shifting legal interpretations owing to specific
historical changes. For instance, just because a law's creators supported
something in the past, say black slavery, that position can be rendered
anachronistic and wholly unacceptable in a changed environment: in this
case, black emancipation.*

*It is important to note that the doctrine of original intent goes hand in
glove with the notion of judicial restraint rather than judicial activism. The
former position confines judicial power to ruling on a measure's constitutionality rather than the actual making of law — seen as the proper province
of the legislature — and the execution of law — the proper province of the
executive branch. Judicial activism, however, tends toward a more historically contingent view of legal decision-making and a more expansive view
of judicial power.*

Philip B. Kurland and Gerhard Casper, eds., *Landmark Briefs and Arguments of the Supreme
Court of the United States: Constitutional Law,* vol. 49 (Arlington, Va.: University Publications of America, 1975), 526–27.

The appellants argued segregated schools violated the lawmakers' original intent in the Fourteenth Amendment; the appellees argued the opposite. This welter of conflicting evidence ultimately impressed the Court as inconclusive. Consequently, the precedential authority of original intent could not be invoked. This section begins with questions the justices asked both sides to answer regarding original intent. These questions also asked for preliminary thoughts about possible remedies. There are three briefs: the appellants', the appellees', and an amicus curiae (friend-of-the-court) brief filed by the federal government for the appellants. The government now openly supported the NAACP's case for school desegregation because of the increasingly powerful understanding that Jim Crow was morally bankrupt, a political liability, and an international embarrassment.

Once the rearguments were heard by the justices, they were better able to fashion a ruling. The result was Brown I: *Chief Justice Earl Warren's March 17, 1954, opinion outlawing Jim Crow and overruling* Plessy v. Ferguson. *This indispensable document, cited in its entirety, concluded this eventful round. Which of the contending responses to each of the justices' questions is more or less persuasive? Why? How does the federal brief compare and contrast with that of the appellants? How do you assess the impact of the federal brief on the appellees' case? How would you characterize the substance and significance of Warren's decision? Do you find the ruling compelling and inspiring? Why or why not?*

. . . 1. What evidence is there that the Congress which submitted and the State legislatures and conventions which ratified the Fourteenth Amendment contemplated or did not contemplate, understood or did not understand, that it would abolish segregation in public schools?

2. If neither the Congress in submitting nor the States in ratifying the Fourteenth Amendment understood that compliance with it would require the immediate abolition of segregation in public schools, was it nevertheless the understanding of the framers of the Amendment

(a) that future Congresses might, in the exercise of their power under section 5 of the Amendment, abolish such segregation, or

(b) that it would be within the judicial power, in light of future conditions, to construe the Amendment as abolishing such segregation of its own force?

3. On the assumption that the answers to questions 2 *(a)* and *(b)* do not dispose of the issue, is it within the judicial power, in construing the Amendment, to abolish segregation in public schools?

4. Assuming it is decided that segregation in public schools violates the Fourteenth Amendment

(a) would a decree necessarily follow providing that, within the limits set by normal geographic school districting, Negro children should forthwith be admitted to schools of their choice, or

(b) may this Court, in the exercise of its equity powers, permit an effective gradual adjustment to be brought about from existing segregated systems to a system not based on color distinctions?

5. On the assumption on which questions 4 *(a)* and *(b)* are based, and assuming further that this Court will exercise its equity powers to the end described in question 4 *(b)*,

(a) should this Court formulate detailed decrees in these cases;

(b) if so, what specific issues should the decrees reach;

(c) should this Court appoint a special master to hear evidence with a view to recommending specific terms for such decrees;

(d) should this Court remand to the courts of first instance with directions to frame decrees in these cases, and if so what general directions should the decrees of this Court include and what procedures should the courts of first instance follow in arriving at the specific terms of more detailed decrees?

The Attorney General of the United States is invited to take part in the oral argument and to file an additional brief if he so desires.

Appellants' Brief

1953

... The substantive question... is whether a state can, consistently with the Constitution, exclude children, solely on the ground that they are Negroes, from public schools which otherwise they would be qualified to attend. It is the thesis of this brief, submitted on behalf of the excluded children, that the answer to the question is in the negative: the Fourteenth Amendment prevents states from according differential treatment to American children on the basis of their color or race. Both the legal precedents and the judicial theories, ... and the evidence concerning the intent of the framers of the Fourteenth Amendment and the under-

Philip B. Kurland and Gerhard Casper, eds., *Landmark Briefs and Arguments of the Supreme Court of the United States: Constitutional Law,* vol. 49 (Arlington, Va.: University Publications of America, 1975), 528–33.

standing of the Congress and the ratifying states ... support this proposition.

Denying this thesis, the school authorities, relying in part on language originating in this Court's opinion in *Plessy* v. *Ferguson,* 163 U.S. 537, urge that the exclusion of Negroes ... from designated public schools is permissible when the excluded children are afforded admittance to other schools especially reserved for Negroes ... if such schools are equal.

The procedural question common to all the cases is the role to be played, and the time-table to be followed, by this Court and the lower courts in directing an end to the challenged exclusion, in the event that this Court determines, with respect to the substantive question, that exclusion of Negroes ... from public schools contravenes the Constitution.

The importance to our American democracy of the substantive question can hardly be overstated. The question is whether a nation founded on the proposition that "all men are created equal" is honoring its commitments to grant "due process of law" and "the equal protection of the laws" to all within its borders when it, or one of its constituent states, confers or denies benefits on the basis of color or race.

1. Distinctions drawn by state authorities on the basis of color or race violate the Fourteenth Amendment. ... This has been held to be true even as to the conduct of public educational institutions. ... Whatever other purposes the Fourteenth Amendment may have had, it is indisputable that its primary purpose was to complete the emancipation provided by the Thirteenth Amendment by ensuring to the Negro equality before the law. ...

2. Even if the Fourteenth Amendment did not *per se* invalidate racial distinctions as a matter of law, the racial segregation challenged in the instant cases would run afoul of the conventional test established for application of the equal protection clause because the racial classifications here have no reasonable relation to any valid legislative purpose. ...

3. Appraisal of the facts requires rejection of the contention of the school authorities. The educational detriment involved in racially constricting a student's associations has already been recognized by this Court. ...

4. The argument that the requirements of the Fourteenth Amendment are met by providing alternative schools rests, finally, on reiteration of the separate but equal doctrine enunciated in *Plessy* v. *Ferguson.*

Were these ordinary cases, it might be enough to say that the *Plessy* case can be distinguished—that it involved only segregation in transportation. But these are not ordinary cases, and in deference to their importance it seems more fitting to meet the *Plessy* doctrine head-on and to declare that doctrine erroneous.

Candor requires recognition that the plain purpose and effect of segregated education is to perpetuate an inferior status for Negroes which is America's sorry heritage from slavery. But the primary purpose of the Fourteenth Amendment was to deprive the states of *all* power to perpetuate such a caste system.

5. The first and second of the five questions propounded by this Court requested enlightenment as to whether the Congress which submitted, and the state legislatures and conventions which ratified, the Fourteenth Amendment contemplated or understood that it would prohibit segregation in public schools, either of its own force or through subsequent legislative or judicial action. The evidence, both in Congress and in the legislatures of the ratifying states, reflects the substantial intent of the Amendment's proponents and the substantial understanding of its opponents that the Fourteenth Amendment would, of its own force, proscribe all forms of state-imposed racial distinctions, thus necessarily including all racial segregation in public education.

The Fourteenth Amendment was actually the culmination of the determined efforts of the Radical Republican majority in Congress to incorporate into our fundamental law the well-defined equalitarian principle of complete equality for all without regard to race or color. The debates in the 39th Congress and succeeding Congresses clearly reveal the intention that the Fourteenth Amendment would work a revolutionary change in our state-federal relationship by denying to the states the power to distinguish on the basis of race.

The Civil Rights Bill of 1866, as originally proposed, possessed scope sufficiently broad in the opinion of many Congressmen to entirely destroy all state legislation based on race. A great majority of the Republican Radicals—who later formulated the Fourteenth Amendment—understood and intended that the Bill would prohibit segregated schools. Opponents of the measure shared this understanding. The scope of this legislation was narrowed because it was known that the Fourteenth Amendment was in process of preparation and would itself have scope exceeding that of the original draft of the Civil Rights Bill.

6. The evidence makes clear that it was the intent of the proponents of the Fourteenth Amendment, and the substantial understanding of its opponents, that it would, of its own force, prohibit all state action predicated upon race or color. The intention of the framers with respect to any specific example of caste state action—in the instant cases, segregated education—cannot be determined solely on the basis of a tabulation of

contemporaneous statements mentioning the specific practice. The framers were formulating a constitutional provision setting broad standards for determination of the relationship of the state to the individual. In the nature of things they could not list all the specific categories of existing and prospective state activity which were to come within the constitutional prohibitions. The broad general purpose of the Amendment—obliteration of race and color distinctions—is clearly established by the evidence. So far as there was consideration of the Amendment's impact upon the undeveloped educational systems then existing, both proponents and opponents of the Amendment understood that it would proscribe all racial segregation in public education.

7. While the Amendment conferred upon Congress the power to enforce its prohibitions, members of the 39th Congress and those of subsequent Congresses made it clear that the framers understood and intended that the Fourteenth Amendment was self-executing and particularly pointed out that the federal judiciary had authority to enforce its prohibitions without Congressional implementation.

8. The evidence as to the understanding of the states is equally convincing. Each of the eleven states that had seceded from the Union ratified the Amendment, and concurrently eliminated racial distinctions from its laws, and adopted a constitution free of requirement or specific authorization of segregated schools. Many rejected proposals for segregated schools, and none enacted a school segregation law until after readmission. The significance of these facts is manifest from the consideration that ten of these states, which were required, as a condition of readmission, to ratify the Amendment and to modify their constitutions and laws in conformity therewith, considered that the Amendment required them to remove all racial distinctions from their existing and prospective laws, including those pertaining to public education.

Twenty-two of the twenty-six Union states also ratified the Amendment. Although unfettered by congressional surveillance, the overwhelming majority of the Union states acted with an understanding that it prohibited racially segregated schools and necessitated conformity of their school laws to secure consistency with that understanding.

9. In short, the historical evidence fully sustains this Court's conclusion in the *Slaughter Houses Cases,* 16 Wall. 61, 81, that the Fourteenth Amendment was designed to take from the states all power to enforce caste or class distinctions.

10. The Court in its fourth and fifth questions assumes that segregation is declared unconstitutional and inquires as to whether relief should be granted immediately or gradually. Appellants, recognizing the possi-

bility of delay of a purely administrative character, do not ask for the impossible. No cogent reasons justifying further exercise of equitable discretion, however, have as yet been produced.

It has been indirectly suggested in the briefs and oral argument of appellees that some such reasons exist. Two plans were suggested by the United States in its Brief as *Amicus Curiae*. We have analyzed each of these plans as well as appellees' briefs and oral argument and find nothing there of sufficient merit on which this Court, in the exercise of its equity power, could predicate a decree permitting an effective gradual adjustment from segregated to non-segregated school systems. Nor have we been able to find any other reasons or plans sufficient to warrant the exercise of such equitable discretion in these cases. Therefore, in the present posture of these cases, appellants are unable to suggest any compelling reasons for this Court to postpone relief.

Appellees' Brief

1953

1. All available evidence points to the conclusion that a majority of the Congress which submitted the Fourteenth Amendment, did not contemplate or understand that it would abolish segregation in the public schools. An analysis of the Congressional debates and the currents of Abolitionist thought during the period prior to and contemporaneous with the adoption of the Fourteenth Amendment indicates that the general concern was with the fundamental rights of life, liberty and property and that by Section 1 of the Fourteenth Amendment they sought to guarantee the negro equality of enjoyment of these rights. The right to mingle with other races in the public schools was not included in this concept of basic rights. The Congress further demonstrated by direct legislation that it did not consider segregation to be a violation of constitutional rights. By its own act it provided segregated schools for the District of Columbia, over which it has exclusive legislative power. Furthermore, subsequent to adoption of the Fourteenth Amendment, the

Philip B. Kurland and Gerhard Casper, eds., *Landmark Briefs and Arguments of the Supreme Court of the United States: Constitutional Law,* vol. 49 (Arlington, Va.: University Publications of America, 1975), 774–75, 817–18.

Congress specifically refused to enact legislation prohibiting segregation in the public schools.

2. Neither was it the understanding of the states which ratified the Fourteenth Amendment that it would abolish segregation in the public schools. The laws of a majority of the states authorized segregation at the time the Fourteenth Amendment was ratified. There is no evidence that any state altered its policy in this respect by reason of the Fourteenth Amendment. On the other hand, it is a fact that in at least ten states the same legislatures that ratified the Fourteenth Amendment enacted legislation providing separate schools. This we deem evidence that the states did not contemplate that segregation was precluded by the Fourteenth Amendment.

3. Inasmuch as it was the understanding of the framers, the Congress which submitted, and the states which ratified the Fourteenth Amendment, that segregation in the public schools was not included within its purview, it could not have been contemplated at that time that any future federal authority, either legislative or judicial, might abolish segregation without further grant of constitutional authority. However, if, in spite of evidence to the contrary, it is conceived that segregation might be brought within the purview of federal authority, then, consistent with the intent of the framers, it is for the Congress and not the judiciary to act.

4. It is not within the judicial powers to construe the Fourteenth Amendment in a manner differently than the framers intended with reference to facts existing at the time the amendment was adopted, where no substantial change in conditions is shown. It was the intent of the framers of the Fourteenth Amendment that the management of the public schools should be left to the state legislatures. The federal judiciary may not now assume jurisdiction over this phase of state activity in disregard of the framers' intent.

5. The people of Kansas, through the normal processes of local government, are demonstrating their willingness and capacity to deal with local race problems in a manner most beneficial to all concerned. Federal interference is neither necessary nor justified. . . .

Finally, we are requested to discuss the characteristics of the decree that should be entered in event the Court should reach the improbable conclusion that racial segregation in the public schools per se is violative of constitutional rights. With respect to these questions, the Appellee Board of Education has filed its separate brief herein, setting out its views in detail and calling attention to certain local conditions that would be affected by such a decree.

Inasmuch as the abolishment of the policy of segregation would require no change in state policy, and in view of the fact that Appellee Board of Education has discussed these questions fully in its separate brief, detailed comment by us is not necessary. We recognize, however, that other communities of Kansas would necessarily be affected by the consequences of a decree abolishing segregation, and for that reason we offer the following general views, based on our knowledge of conditions prevailing in Kansas:

We do not believe that the reversal of the judgment of the court below would necessarily require admission of Negro children forthwith to the schools of their choice. On the contrary, we believe that the Court in the exercise of its equity powers may permit an effective gradual adjustment. We think that the necessity for safeguarding the integrity of the school system at large must be reconciled with the necessity of effecting constitutional guaranties. We feel that any order that might currently disrupt the orderly conduct of the public school system would not be consistent with a proper exercise of equity jurisdiction.

It is our further view that this Court should not undertake to formulate a detailed decree in this case. Should it deem the judgment of the court below to be improper, we suggest that that judgment should simply be reversed and the cause remanded to that court with directions to frame an appropriate decree. In the preparation of such a decree the District Court might properly conduct such investigations and hearings as are necessary to apprise itself of the various interests to be reconciled. However, that is a problem that should be determined on the local level by the court of original jurisdiction. We can assure this Court that the State of Kansas and its local boards of education will act in complete good faith to comply with the letter and the spirit of any decree that may be entered.

Federal Friend-of-the-Court Brief
1953

... The Congressional history of the Fourteenth Amendment shows that the Amendment was proposed and debated as part of a broad and continuing program to establish full freedom and legal equality for Negroes. Many in the Congress which considered the Thirteenth Amendment understood it to abolish not only slavery but also its concomitant legal discriminations. This understanding rested on a belief that that Amendment had made the Negro an indistinguishable part of the population and hence entitled to the same rights and privileges under the laws as all others. The enactment of the Black Codes in the Southern states made it obvious, however, that additional protection by the national government was required.

The civil rights legislation enacted by the 39th Congress was designed to strike down distinctions based on race or color. From the debates on that legislation, however, there emerged the view that the Thirteenth Amendment alone did not afford a sufficient constitutional basis for such action, and that a further amendment was necessary. In the same debates there was also crystallized the view that only explicit constitutional embodiment of the principle of equality before the law could protect that principle from change by some future Congress. ...

Neither the majority nor the minority in the 39th Congress evidenced any substantial disagreement as to the broad scope of Section 1 of the Amendment. The majority repeatedly affirmed that it would firmly secure the principle that the "law which operates on one man shall operate equally upon all" and would prohibit all legislation by the states drawn on the basis of race and color. The opposition similarly understood its broad purpose; it was on that basis that they voiced their objections.

While the debates reflect a clear understanding as to the breadth of the principle of equality under law embodied in the Fourteenth Amendment, neither its proponents nor its opponents found it necessary or appropriate to catalog exhaustively the specific application of its general principle. Only a few such examples were given during the debates on the Amendment itself. It is noteworthy that one of the majority spokes-

Philip B. Kurland and Gerhard Casper, eds., *Landmark Briefs and Arguments of the Supreme Court of the United States: Constitutional Law,* vol. 49 (Arlington, Va.: University Publications of America, 1975), 978–84, 991–92, 995, 1053–54.

men, at a time when the majority was proceeding under the discipline of party caucus, illustrated the racial discriminations which the Amendment would reach by reference to a state law discriminating against Negroes in public schools. He did not, however, make specific mention of the system of racial segregation which the state law required.

In the debates on the civil rights legislation, which are an integral part of the immediate background of the Fourteenth Amendment, the minority expressed the view that existing state systems of racially-segregated public schools would be stricken down by the broad principle of equal treatment under the law. This view was not disputed by the majority. . . .

In sum, while the legislative history does not conclusively establish that the Congress which proposed the Fourteenth Amendment specifically understood that it would abolish racial segregation in the public schools, there is ample evidence that it did understand that the Amendment established the broad constitutional principle of full and complete equality of all persons under the law, and that it forbade all legal distinctions based on race or color. Concerned as they were with securing to the Negro freedmen these fundamental rights of liberty and equality, the members of Congress did not pause to enumerate in detail all the specific applications of the basic principle which the Amendment incorporated into the Constitution. There is some evidence that this broad principle was understood to apply to racial discriminations in education and that it might have the additional effect of invalidating state laws providing for racial segregation in the public schools.

There is a paucity of available evidence as to the understanding of the state legislatures which ratified the Amendment, in part because of the almost complete absence of records of debates, in part perhaps because their function was to accept or reject a proposal rather than to draft one.

In the states most attention was given to the political aspects of the Republican "plan of reconstruction," which received popular approval in the elections of 1866. It was frequently stated that the Amendment guaranteed to the Negroes full rights of equality as citizens, but the scope and content of those rights were not detailed. The opponents of the Amendment objected to the first section on the ground that it, together with the fifth section, expanded the powers of the Federal Government at the expense of the rights of the states. There were almost no references to schools during consideration of the amendment.

At the time of consideration and ratification of the Fourteenth Amendment, some of the Northern states had and continued segregated schools and some of the Southern states, in providing for the first time for public education for Negroes, established separate schools. In the historical context in which these actions were taken, however, they do not evidence an understanding as to the reach of the Fourteenth Amendment. The inferences to be drawn from these actions necessarily rest on conjecture and speculation. The scanty evidence available suggests that the legislatures were probably unaware that the Amendment was relevant to education, even to the extent of requiring equal, though separate, schools. Proponents of education for Negroes based their arguments on grounds other than the Fourteenth Amendment, and made no reference to it.

In sum, the available materials are too sparse, and the specific references to education too few, to justify any definite conclusion that the state legislatures which ratified the Fourteenth Amendment understood either that it permitted or that it prohibited separate schools.

There is no direct evidence at the time of the adoption of the Amendment that its framers understood specifically that future Congresses might, in the exercise of their power under section 5, abolish segregation in the public schools. They clearly understood, however, that Congress would have the power to enforce the broad guarantees of the Amendment, and the Amendment was deliberately framed so as to assure that the rights protected by section 1 could not be withdrawn or restricted by future Congresses.

Subsequently, in the debates on the Civil Rights Act of 1875, some of the framers expressed an understanding that segregated schools were contrary to the Amendment and that Congress could and should abolish them. While an express prohibition against segregated schools was not contained in the Act in its final form, its omission did not spring from doubt of the power of Congress to enact such a prohibition. . . .

. . . [T]he opinions of this Court, extending over a period of more than three-quarters of a century, . . . show a consistent recognition that the Fourteenth Amendment is to be construed liberally so as to carry out the great and pervading purpose of its framers to establish complete equality for Negroes in the enjoyment of fundamental human rights and to secure those rights against enforcement of legal distinctions based on race or color. . . .

As has been shown, no conclusive evidence of a specific understanding as to the effect of the Fourteenth Amendment on school segregation has been found in its legislative history. But this Court has neither declared nor applied any canon of constitutional interpretation that a construction of an amendment which is warranted by its provisions and manifest policy cannot be adopted unless it is also affirmatively supported by specific evidence in the legislative history showing that its framers so "intended." ... To be sure, the Court will review "the background and environment" of the period in order to illuminate the broad purposes which an amendment was designed to achieve. ... In attempting to determine the application of the amendment to a specific issue, however, the Court will give scant regard to inconclusive excerpts from debates which are relied upon to show a "legislative intent." ... The opinions of the Court, particularly those which have come to be recognized as landmarks in the development of American constitutional law, are replete with expressions of a similar nature. ...

In response to the questions stated in the Court's order directing reargument of these cases, the United States respectfully submits ... (4) that it is within the judicial power to direct such relief as will be effective and just in eliminating existing segregated school systems; and (5) that if the Court holds that laws providing for separate public schools for white and colored children are unconstitutional, it should remand the instant cases to the lower courts with directions to carry out the Court's decision as expeditiously as the particular circumstances permit, as indicated *supra*.

CHIEF JUSTICE EARL WARREN

Opinion of the Court in Brown v. Board of Education

May 17, 1954

These cases come to us from the States of Kansas, South Carolina, Virginia, and Delaware. They are premised on different facts and different local conditions, but a common legal question justifies their consideration together in this consolidated opinion.[1]

[1] In the Kansas case, *Brown* v. *Board of Education,* the plaintiffs are Negro children of elementary school age residing in Topeka. They brought this action in the United States District Court for the District of Kansas to enjoin enforcement of a Kansas statute which permits, but does not require, cities of more than 15,000 population to maintain separate school facilities for Negro and white students. Kan. Gen. Stat. § 72-1724 (1949). Pursuant to that authority, the Topeka Board of Education elected to establish segregated elementary schools. Other public schools in the community, however, are operated on a nonsegregated basis. The three-judge District Court, convened under 28 U. S. C. §§ 2281 and 2284, found that segregation in public education has a detrimental effect upon Negro children, but denied relief on the ground that the Negro and white schools were substantially equal

Leon Friedman, ed., *Argument: The Oral Argument before the Supreme Court in Brown v. Board of Education of Topeka, 1952–1955* (New York: Chelsea House, 1969), 325–31.

In each of the cases, minors of the Negro race, through their legal representatives, seek the aid of the courts in obtaining admission to the public schools of their community on a nonsegregated basis. In each instance, they had been denied admission to schools attended by white children under laws requiring or permitting segregation according to race. This segregation was alleged to deprive the plaintiffs of the equal protection of the laws under the Fourteenth Amendment. In each of the cases other than the Delaware case, a three-judge federal district court denied relief

---

with respect to buildings, transportation, curricula, and educational qualifications of teachers. 98 F. Supp. 797. The case is here on direct appeal under 28 U. S. C. § 1253.

In the South Carolina case, *Briggs* v. *Elliott,* the plaintiffs are Negro children of both elementary and high school age residing in Clarendon County. They brought this action in the United States District Court for the Eastern District of South Carolina to enjoin enforcement of provisions in the state constitution and statutory code which require the segregation of Negroes and whites in public schools. S.C. Const., Art. XI, § 7; S.C. Code § 5377 (1942). The three-judge District court, convened under 28 U. S. C. §§ 2281 and 2284, denied the requested relief. The court found that the Negro schools were inferior to the white schools and ordered the defendants to begin immediately to equalize the facilities. But the court sustained the validity of the contested provisions and denied the plaintiffs admission to the white schools during the equalization program. 98 F. Supp. 529. This Court vacated the District Court's judgment and remanded the case for the purpose of obtaining the court's views on a report filed by the defendants concerning the progress made in the equalization program. 342 U.S. 350. On remand, the District Court found that substantial equality had been achieved except for buildings and that the defendants were proceeding to rectify this inequality as well. 103 F. Supp. 920. The case is again here on direct appeal under 28 U. S. C. § 1253.

In the Virginia case, *Davis* v. *County School Board,* the plaintiffs are Negro children of high school age residing in Prince Edward County. They brought this action in the United States District Court for the Eastern District of Virginia to enjoin enforcement of provisions in the state constitution and statutory code which require the segregation of Negroes and whites in public schools. Va. Const., § 140; Va. Code § 22-221 (1950). The three-judge District Court, convened under 28 U. S. C. §§ 2281 and 2284, denied the requested relief. The court found the Negro school inferior in physical plant, curricula, and transportation, and ordered the defendants forthwith to provide substantially equal curricula and transportation and to "proceed with all reasonable diligence and dispatch to remove" the inequality in physical plant. But, as in the South Carolina case, the court sustained the validity of the contested provisions and denied the plaintiffs admission to the white schools during the equalization program. 103 F. Supp. 337. The case is here on direct appeal under 28 U. S. C. § 1253.

In the Delaware case, *Gebhart* v. *Belton,* the plaintiffs are Negro children of both elementary and high school age residing in New Castle County. They brought this action in the Delaware Court of Chancery to enjoin enforcement of provisions in the state constitution and statutory code which require the segregation of Negroes and whites in public schools. De. Const., Art. X, § 2; Del Rev. Code § 2631 (1935). The Chancellor gave judgment for the plaintiffs and ordered their immediate admission to schools previously attended only by white children, on the ground that the Negro schools were inferior with respect to teacher training, pupil-teacher ratio, extracurricular activities, physical plant, and time and distance involved in travel. 87 A. 2d 862. The Chancellor also found that segregation itself results in an inferior education for Negro children (see note 10, *infra*), but did not rest his decision on that ground. *Id.,* at 865. The Chancellor's decree was affirmed by the Supreme

to the plaintiffs on the so-called "separate but equal" doctrine announced by this Court in *Plessy* v. *Ferguson,* 163 U.S. 537. Under that doctrine, equality of treatment is accorded when the races are provided substantially equal facilities, even though these facilities be separate. In the Delaware case, the Supreme Court of Delaware adhered to that doctrine, but ordered that the plaintiffs be admitted to the white schools because of their superiority to the Negro schools.

The plaintiffs contend that segregated public schools are not "equal" and cannot be made "equal," and that hence they are deprived of the equal protection of the laws. Because of the obvious importance of the question presented, the Court took jurisdiction.[2] Argument was heard in the 1952 Term, and reargument was heard this Term on certain questions propounded by the Court.[3]

Reargument was largely devoted to the circumstances surrounding the adoption of the Fourteenth Amendment in 1868. It covered exhaustively consideration of the Amendment in Congress, ratification by the states, then existing practices in racial segregation, and the views of proponents and opponents of the Amendment. This discussion and our own investigation convince us that, although these sources cast some light, it is not enough to resolve the problem with which we are faced. At best, they are inconclusive. The most avid proponents of the post-War Amendments undoubtedly intended them to remove all legal distinctions among "all persons born or naturalized in the United States." Their opponents, just as certainly, were antagonistic to both the letter and the spirit of the Amendments and wished them to have the most limited effect. What others in Congress and the state legislatures had in mind cannot be determined with any degree of certainty.

An additional reason for the inconclusive nature of the Amendment's history, with respect to segregated schools, is the status of public education at that time.[4] In the South, the movement toward free common

---

Court of Delaware, which intimated, however, that the defendants might be able to obtain a modification of the decree after equalization of the Negro and white schools had been accomplished. 91 A. 2d 137, 152. The defendants, contending only that the Delaware courts had erred in ordering the immediate admission of the Negro plaintiffs to the white schools, applied to this Court for certiorari. The writ was granted, 344 U. S. 891. The plaintiffs, who were successful below, did not submit a cross-petition.

[2]344 U.S. 1, 141, 891.

[3]345 U.S. 972. The Attorney General of the United States participated both Terms as *amicus curiae.*

[4]For a general study of the development of public education prior to the Amendment, see Butts and Cremin, A History of Education in American Culture (1953), Pts. I, II; Cubberley, Public Education in the United States (1934 ed.), cc. II-XII. School practices current at the time of the adoption of the Fourteenth Amendment are described in Butts and

schools, supported by general taxation, had not yet taken hold. Education of white children was largely in the hands of private groups. Education of Negroes was almost nonexistent, and practically all of the race were illiterate. In fact, any education of Negroes was forbidden by law in some states. Today, in contrast, many Negroes have achieved outstanding success in the arts and sciences as well as in the business and professional world. It is true that public school education at the time of the Amendment had advanced further in the North, but the effect of the Amendment on Northern States was generally ignored in the congressional debates. Even in the North, the conditions of public education did not approximate those existing today. The curriculum was usually rudimentary; ungraded schools were common in rural areas; the school term was but three months a year in many states; and compulsory school attendance was virtually unknown. As a consequence, it is not surprising that there should be so little in the history of the Fourteenth Amendment relating to its intended effect on public education.

In the first cases in this Court construing the Fourteenth Amendment, decided shortly after its adoption, the Court interpreted it as proscribing all state-imposed discriminations against the Negro race.[5] The doctrine of "separate but equal" did not make its appearance in this Court until

---

Cremin, *supra,* at 269–275; Cubberley, *supra,* at 288–339, 408–431; Knight, Public Education in the South (1922), cc. VIII, IX. See also H. Ex. Doc. No. 315, 41st Cong., 2d Sess. (1871). Although the demand for free public schools followed substantially the same pattern in both the North and the South, the development in the South did not begin to gain momentum until about 1850, some twenty years after that in the North. The reasons for the somewhat slower development in the South (*e.g.,* the rural character of the South and the different regional attitudes toward state assistance) are well explained in Cubberley, *supra,* at 408–423. In the country as a whole, but particularly in the South, the War virtually stopped all progress in public education. *Id.,* at 427–428. The low status of Negro education in all sections of the country, both before and immediately after the War, is described in Beale, A History of Freedom of Teaching in American Schools (1941), 112–132, 175–195. Compulsory school attendance laws were not generally adopted until after the ratification of the Fourteenth Amendment, and it was not until 1918 that such laws were in force in all the states. Cubberley, *supra,* at 563–565.

[5] *Slaughter-House Cases,* 16 Wall. 36, 67–72 (1873); *Strauder* v. *West Virginia,* 100 U.S. 303, 307–308 (1880): "It ordains that no State shall deprive any person of life, liberty, or property, without due process of law, or deny to any person within its jurisdiction the equal protection of the laws. What is this but declaring that the law in the States shall be the same for the black as for the white; that all persons, whether colored or white, shall stand equal before the laws of the States, and, in regard to the colored race, for whose protection the amendment was primarily designed, that no discrimination shall be made against them by law because of their color? The words of the amendment, it is true, are prohibitory, but they contain a necessary implication of a positive immunity, or right, most valuable to the colored race, — the right to exemption from unfriendly legislation against them distinctively as colored, — exemption from legal discriminations, implying inferiority in civil society, lessening the security of their enjoyment of the rights which others enjoy, and discriminations

1896 in the case of *Plessy* v. *Ferguson, supra,* involving not education but transportation.[6] American courts have since labored with the doctrine for over half a century. In this Court, there have been six cases involving the "separate but equal" doctrine in the field of public education.[7] In *Cumming* v. *County Board of Education,* 175 U.S. 528, and *Gong Lum* v. *Rice,* 275 U.S. 78, the validity of the doctrine itself was not challenged.[8] In more recent cases, all on the graduate school level, inequality was found in that specific benefits enjoyed by white students were denied to Negro students of the same educational qualifications. *Missouri ex rel. Gaines* v. *Canada,* 305 U.S. 337; *Sipuel* v. *Oklahoma,* 332 U.S. 631; *Sweatt* v. *Painter,* 339 U.S. 629; *McLaurin* v. *Oklahoma State Regents,* 339 U.S. 637. In none of these cases was it necessary to re-examine the doctrine to grant relief to the Negro plaintiff. And in *Sweatt* v. *Painter, supra,* the Court expressly reserved decision on the question whether *Plessy* v. *Ferguson* should be held inapplicable to public education.

In the instant cases, that question is directly presented. Here, unlike *Sweatt* v. *Painter,* there are findings below that the Negro and white schools involved have been equalized, or are being equalized, with respect to buildings, curricula, qualifications and salaries of teachers, and other "tangible" factors.[9] Our decision, therefore, cannot turn on merely a comparison of these tangible factors in the Negro and white schools involved in each of the cases. We must look instead to the effect of segregation itself on public education.

---

which are steps towards reducing them to the condition of a subject race." See also *Virginia* v. *Rives,* 100 U.S. 313, 318 (1880); *Ex parte Virginia,* 100 U.S. 339, 344–345 (1880).

[6]The doctrine apparently originated in *Roberts* v. *City of Boston,* 59 Mass. 198, 206 (1850), upholding school segregation against attack as being violative of a state constitutional guarantee of equality. Segregation in Boston public schools was eliminated in 1855. Mass. Acts 1855, c. 256. But elsewhere in the North segregation in public education has persisted in some communities until recent years. It is apparent that such segregation has long been a nationwide problem, not merely one of sectional concern.

[7]See also *Berea College* v. *Kentucky,* 211 U.S. 45 (1908).

[8]In the *Cumming* case, Negro taxpayers sought an injunction requiring the defendant school board to discontinue the operation of a high school for white children until the board resumed operation of a high school for Negro children. Similarly, in the *Gong Lum* case, the plaintiff, a child of Chinese descent, contended only that state authorities had misapplied the doctrine by classifying him [The child was female — ED.] with Negro children and requiring him to attend a Negro school.

[9]In the Kansas case, the court below found substantial equality as to all such factors. 98 F. Supp. 797, 798. In the South Carolina case, the court below found that the defendants were proceeding "promptly and in good faith to comply with the court's decree." 103 F. Supp. 920, 921. In the Virginia case, the court below noted that the equalization program was already "afoot and progressing" (103 F. Supp. 337, 341); since then, we have been advised, in the Virginia Attorney General's brief on reargument, that the program has now been com-

In approaching this problem, we cannot turn the clock back to 1868 when the Amendment was adopted, or even to 1896 when *Plessy* v. *Ferguson* was written. We must consider public education in the light of its full development and its present place in American life throughout the Nation. Only in this way can it be determined if segregation in public schools deprives these plaintiffs of the equal protection of the laws.

Today, education is perhaps the most important function of state and local governments. Compulsory school attendance laws and the great expenditures for education both demonstrate our recognition of the importance of education to our democratic society. It is required in the performance of our most basic public responsibilities, even service in the armed forces. It is the very foundation of good citizenship. Today it is a principal instrument in awakening the child to cultural values, in preparing him for later professional training, and in helping him to adjust normally to his environment. In these days, it is doubtful that any child may reasonably be expected to succeed in life if he is denied the opportunity of an education. Such an opportunity, where the state has undertaken to provide it, is a right which must be made available to all on equal terms.

We come then to the question presented: Does segregation of children in public schools solely on the basis of race, even though the physical facilities and other "tangible" factors may be equal, deprive the children of the minority group of equal educational opportunities? We believe that it does.

In *Sweatt* v. *Painter, supra,* in finding that a segregated law school for Negroes could not provide them equal educational opportunities, this Court relied in large part on "those qualities which are incapable of objective measurement but which make for greatness in a law school." In *McLaurin* v. *Oklahoma State Regents, supra,* the Court, in requiring that a Negro admitted to a white graduate school be treated like all other students, again resorted to intangible considerations: " . . . his ability to study, to engage in discussions and exchange views with other students, and, in general, to learn his profession." Such considerations apply with added force to children in grade and high schools. To separate them from others of similar age and qualifications solely because of their race generates a feeling of inferiority as to their status in the community that may affect their hearts and minds in a way unlikely ever to be undone. The

---

pleted. In the Delaware case, the court below similarly noted that the state's equalization program was well under way. 91 A. 2d 137, 149.

effect of this separation on their educational opportunities was well stated by a finding in the Kansas case by a court which nevertheless felt compelled to rule against the Negro plaintiffs:

> Segregation of white and colored children in public schools has a detrimental effect upon the colored children. The impact is greater when it has the sanction of the law; for the policy of separating the races is usually interpreted as denoting the inferiority of the negro group. A sense of inferiority affects the motivation of a child to learn. Segregation with the sanction of law, therefore, has a tendency to [retard] the educational and mental development of negro children and to deprive them of some of the benefits they would receive in a racial[ly] integrated school system.[10]

Whatever may have been the extent of psychological knowledge at the time of *Plessy* v. *Ferguson,* this finding is amply supported by modern authority.[11] Any language in *Plessy* v. *Ferguson* contrary to this finding is rejected.

We conclude that in the field of public education the doctrine of "separate but equal" has no place. Separate educational facilities are inherently unequal. Therefore, we hold that the plaintiffs and others similarly situated for whom the actions have been brought are, by reason of the segregation complained of, deprived of the equal protection of the laws guaranteed by the Fourteenth Amendment. This disposition makes unnecessary any discussion whether such segregation also violates the Due Process Clause of the Fourteenth Amendment.[12]. . .

[10]A similar finding was made in the Delaware case: "I conclude from the testimony that in our Delaware society, State-imposed segregation in education itself results in the Negro children, as a class, receiving educational opportunities which are substantially inferior to those available to white children otherwise similarly situated." 87 A. 2d 862, 865.

[11]K.B. Clark, Effect of Prejudice and Discrimination on Personality Development (Midcentury White House Conference on Children and Youth, 1950); Witmer and Kotinsky, Personality in the Making (1952), c. VI; Deutscher and Chein, The Psychological Effects of Enforced Segregation: A Survey of Social Science Opinion, 26 J. Psychol. 259 (1948); Chein, What Are the Psychological Effects of Segregation Under Conditions of Equal Facilities?, 3 Int. J. Opinion and Attitude Res. 229 (1949); Brameld, Educational Costs, in Discrimination and National Welfare (MacIver, ed., 1949), 44–48; Frazier, The Negro in the United States (1949), 674–681. And see generally Myrdal, An American Dilemma (1944).

[12]See *Bolling* v. *Sharpe, post,* p. 497, concerning the Due Process Clause of the Fifth Amendment.

**Round Three: Reargument on Remedy —
Immediate or Gradual?**

Appellants' Brief
1954

Round three treated the extremely difficult issue of remedy: Were Jim Crow schools to be dismantled and, as a consequence, integrated schools established expeditiously or incrementally? As evidenced by what we now know of the justices' private concerns, this conundrum haunted them from the beginning of the case. Early on, the Court clearly leaned in the direction of overturning Plessy, but was paralyzed around how to rule on relief without creating a political furor, especially without unduly antagonizing Jim Crow's supporters. The compromise judgment in Brown II — integration with "all deliberate speed" — proved sufficiently flexible, and thus ambiguous enough, to accommodate the justices' varied concerns.

This section includes the appellants' brief for immediate relief, the appellees' brief for gradual relief, and the appellants' reply brief. Chief Justice Warren's compromise and evasive relief decree articulating "all deliberate speed" concludes the section. Which position do you find more cogent: immediatism or gradualism? What do you see as the strengths and weaknesses of Brown II's mandate? How do you assess the impact of popular opinion on one hand and lackluster presidential and legislative leadership on the other on the high court's decision?

I

Answering Question 4: Only a Decree Requiring Desegregation as Quickly as Prerequisite Administrative and Mechanical Procedures Can Be Completed Will Discharge Judicial Responsibil-

Philip B. Kurland and Gerhard Casper, eds., *Landmark Briefs and Arguments of the Supreme Court of the United States: Constitutional Law,* vol. 49A (Arlington, Va.: University Publications of America, 1975), 646–47, 650–67.

ity for the Vindication of the Constitutional Rights of Which Appellants Are Being Deprived.

In the normal course of judicial procedure, this Court's decision that racial segregation in public education is unconstitutional would be effectuated by decrees forthwith enjoining the continuation of that segregation. Indeed, in *Sipuel* v. *Board of Regents* ... when effort was made to secure postponement of the enforcement of similar rights, this Court not only refused to delay action but accelerated the granting of relief by ordering its mandate to issue forthwith.

In practical effect, such disposition of this litigation would require immediate initiation of the administrative procedures prerequisite to desegregation, to be followed by the admission of the complaining children and others similarly situated to unsegregated schools at the beginning of the next academic term. This means that appellees will have had from May 17, 1954, to September, 1955, to complete whatever adjustments may be necessary.

If appellees desire any postponement of relief beyond that date, the affirmative burden must be on them to state explicitly what they propose and to establish that the requested postponement has judicially cognizable advantages greater than those inherent in the prompt vindication of appellants' adjudicated constitutional rights. Moreover, when appellees seek to postpone the enjoyment of rights which are personal and present ... that burden is particularly heavy. When the rights of school children are involved the burden is even greater. Each day relief is postponed is to the appellants ... a day of serious and irreparable injury; for this Court has announced that segregation of Negroes in the public schools "generates a feeling of inferiority as to their status in the community that may affect their hearts and minds in a way unlikely ever to be undone. . . . " And, time is of the essence because the period of public school attendance is short.

A. AGGRIEVED PARTIES SHOWING DENIAL OF CONSTITUTIONAL
RIGHTS IN ANALOGOUS SITUATIONS HAVE RECEIVED
IMMEDIATE RELIEF DESPITE ARGUMENTS FOR DELAY
MORE PERSUASIVE THAN ANY AVAILABLE HERE.

Where a substantial constitutional right would be impaired by delay, this Court has refused to postpone injunctive relief even in the face of the gravest of public considerations. . . .

Counsel have discovered no case wherein this Court has found a violation of a present constitutional right but has postponed relief on the representation by governmental officials that local mores and customs justify delay which might produce a more orderly transition.

It would be paradoxical indeed if, in the instant cases, it were decided for the first time that constitutional rights may be postponed because of anticipation of difficulties arising out of local feelings. These cases are brought to vindicate rights which, as a matter of common knowledge and legal experience, need, above all others, protection against local attitudes and patterns of behavior.[1] They are brought, specifically, to uphold rights under the Fourteenth Amendment which are not to be qualified, substantively or remedially, by reference to local mores. On the contrary, the Fourteenth Amendment, on its face and as a matter of history, was designed for the very purpose of affording protection against local mores and customs, and Congress has implemented that design by providing redress against aggression under color of state laws, customs and usages. . . .

Surely, appellants' rights are not to be enforced at a pace geared down to the very customs which the Fourteenth Amendment and implementing federal laws were designed to combat. . . .

It should be remembered that the rights involved in these cases are not only of importance to appellants and the class they represent, but are among the most important in our society.

B. Empirical Data Negate Unsupported Speculations That a Gradual Decree Would Bring About a More Effective Adjustment. . . . There is no basis for the assumption that gradual as opposed to immediate desegregation is the better, smoother or more "effective" mode of transition. On the contrary, there is an impressive body of evidence which supports the position that gradualism, far from facilitating the process, may actually make it more difficult; that in fact, the problems of transition will be a good deal less complicated than might be forecast by appellees. Our submission is that this, like many wrongs, can be easiest and best undone, not by "tapering off" but by forthright action. . . .

Some plans have been tried involving a set "deadline" without the specification of intervening steps to be taken. Where such plans have been tried, the tendency seems to have been to regard the deadline as the time when action is to be initiated rather than the time at which desegregation is to be accomplished. Since there exists no body of knowledge that is even helpful in selecting an optimum time at the end

---

[1] In the instant cases, dark and uncertain prophecies as to anticipated community reactions to school desegregation are speculative at best.

of which the situation may be expected to be better, the deadline date is necessarily arbitrary and hence may be needlessly remote.[2]

A species of the "deadline" type of plan attempts to prepare the public, through churches, radio and other agencies, for the impending change. It is altogether conjectural how successful such attempts might be in actually effecting change in attitude. The underlying assumption — that change in attitude must precede change in action — is itself at best a highly questionable one. There is a considerable body of evidence to indicate that attitude may itself be influenced by situation[3] and that, where the situation demands that an individual act as if he were not prejudiced, he will so act, despite the continuance, at least temporarily, of the prejudice.[4] We submit that this Court can itself contribute to an effective and decisive change in attitude by insistence that the present unlawful situation be changed forthwith.

As to any sort of "deadline" plan, even assuming that community leaders make every effort to build community support for desegregation, experience shows that other forces in the community will use the time allowed to firm up and build opposition.[5] At least in South Carolina and Virginia, as well as in some other states affected by this decision, statements and action of governmental officials since May 17th demonstrate that they will not use the time allowed to build up community support for

[2]Ashmore, *op. cit. supra* note 6, at 70, 71, 79, 80; Clark, *op. cit. supra* note 6, at 36, 45.

[3]Clark, *op. cit. supra* note 6, at 69–76.

[4]Kutner, Wilkins and Yarrow, Verbal Attitudes and Overt Behavior Involving Racial Prejudice, 47 J. Abnormal and Social Psych. 649–652 (1952); La Piere, Attitudes vs. Action, 13 Social Forces 230–237 (1934); Saenger and Gilbert, Customer Reactions to the Integration of Negro Sales Personnel, 4 Int. J. Opinion and Attitudes Research 57–76 (1950); Deutsch and Collins, Interracial Housing, A Psychological Study of a Social Experiment (1951); Chein, Deutsch, Hyman and Jahoda, Consistency and Inconsistency in Intergroup Relations, 5 J. Social Issues 1–63 (1949). Ashmore, *op. cit. supra* note 6, at 42; New York Times, "Mixed Schools Set in 'Border' States," August 29, 1954, p. 88, col. 1–4; New York Times, "New Mexico Town Quietly Ends Pupil Segregation Despite a Cleric," August 31, 1954, p. 1, col. 3–4; Rose, You Can't Legislate Against Prejudice — Or Can You?, 9 Common Ground 61–67 (1949), reprinted in Race Prejudice and Discrimination, (Rose ed. 1951); Nichols, Breakthrough on the Color Front (1954); Merton, West and Jahoda, Social Fictions and Social Facts: The Dynamics of Race Relations in Hilltown, Columbia University Bureau of Applied Social Research (mimeographed); Merton, West, Jahoda and Selden, Social Policy and Social Research in Housing, 7 J. Social Issues, 132–140 (1951); Merton, The Social Psychology of Housing (1948).

South as well as North, people's actions and attitudes were changed not in advance of but after the admission of Negroes into organized baseball. See Clement, Racial Integration in the Field of Sports, 23 J. Negro Ed. 226–228 (1954). Objections to desegregation have generally been found to be greater before than after its accomplishment. Clark, *op. cit. supra* note 6, *passim;* Conference Report, Arizona Council for Civic Unity Conference on School Segregation (Phoenix, Arizona, June 2, 1951).

[5]Clark, *op cit. supra* note 6, at 43, 44; Brogan, The Emerson School — Community Problem, Gary, Indiana, Bureau of Intercultural Education Report (October 1947, mimeographed); Tipton, Community in Crisis 15–76 (1953).

desegregation.[6] Church groups and others in the South who are seeking to win community acceptance for the Court's May 17th decision cannot be effective without the support of a forthwith decree from this Court.

Besides the "deadline" plans, various "piecemeal" schemes have been suggested and tried. These seem to be inspired by the assumption that it is always easier and better to do something slowly and a little at a time than to do it all at once. As might be expected, it has appeared that the resistance of some people affected by such schemes is increased since they feel arbitrarily selected as experimental animals. Other members in the community observe this reaction and in turn their anxieties are sharpened.[7]

Piecemeal desegregation of schools, on a class-by-class basis, tends to arouse feelings of the same kind[8] and these feelings are heightened by the intra-familial and intra-school differences thus created.[9] It would be hard to imagine any means better calculated to increase tension in regard to desegregation than to so arrange matters so that some children in a family were attending segregated and others unsegregated classes. Hardly more promising of harmony is the prospect of a school which is segregated in the upper parts and mixed in the lower.

When one looks at various "gradual" processes, the fact is that there is no convincing evidence which supports the theory that "gradual" desegregation is more "effective."[10] On the contrary, there is considerable evidence that the forthright way is a most effective way.[11]

[6] For the latest example of this, see New York Times, "7 of South's Governors Warn of 'Dissensions' in Curb on Bias — Avow Right of States to Control Public School Procedures — Six at Meeting Refrain from Signing Statement," November 14, 1954, p. 58, col. 4–5.

[7] Tipton, *op. cit. supra* note 11, at 42, 47, 57, 71; Clark, Some Principles Related to the Problem of Desegregation, 23 J. Negro Ed. 343 (1954); Culver, Racial Desegregation in Education in Indiana, 23 J. Negro Ed. 300 (1954).

[8] Ashmore, *op. cit. supra* note 6, at 79, 80; Clark, Desegregation: An Appraisal of the Evidence, *op. cit. supra* note 6, at 36, 45.

[9] Clark, Effects of Prejudice and Discrimination on Personality Developments, Mid-Century White House Conference on Children and Youth (mimeographed, 1950).

[10] Ashmore, *op. cit. supra* note 6, at 80:

> Proponents of the gradual approach argue that it minimizes public resistance to integration. But some school officials who have experienced it believe the reverse is true. A markedly gradual program, they contend, particularly one which involves the continued maintenance of some separate schools, invites opposition and allows time for it to be organized. Whatever the merit of this argument, the case histories clearly indicate a tendency for local political pressure to be applied by both sides when the question of integration is raised, and when policies remain unsettled for a protracted period the pressures mount. One school board member in Arizona privately expressed the wish that the state had gone all the way and made integration mandatory instead of optional — thus giving the board something to point to as justification for its action.

[11] Clark, *op. cit. supra* note 6, at 46, 47; Wright, Racial Integration in the Public Schools of New Jersey, 23 J. Negro Ed. 283 (1954); Knox, Racial Integration in the Schools of Ari-

The progress of desegregation in the Topeka schools is an example of gradualism based upon conjecture, fears and speculation regarding community opposition which might delay completion of desegregation forever. The desegregation plan adopted by the Topeka school authorities called for school desegregation first in the better residential areas of the city and desegregation followed in those areas where the smallest number of Negro children lived. There is little excuse for the school board's not having already completed desegregation. Apparently either the fact that the school board, in order to complete the transition, may have to utilize one or more of the former schools for Negroes and assign white children to them or the fact that it must now reassign some 700 Negro children to approximately seven former all-white schools, seems to present difficulties to appellees. One must remember that in Topeka there has been complete integration above the sixth grade for many years. The schools already desegregated have reported no difficulties. There can hardly be any basic resistance to nonsegregated schools in the habits or customs of the city's populace. The elimination of the remnants of segregation throughout the city's school system should be a simple matter.

No special public preparations involving teachers, parents, students or the general public were made, nor were they necessary in advance of either the first or second step in the implementation of the Board's decision to desegregate the school system. Indeed, the Board of Education adopted the second step in January, 1954, and the only reports of what was involved were those published in the newspapers. Negro parents living in these territories were not notified by appellees regarding the change, but transferred their children to the schools in question on the basis of information provided in the newspapers. As far as the teachers in those schools were concerned, they were merely informed in the Spring of 1954 that their particular schools would be integrated in September. Thus, delay here cannot be based upon need for public orientation.

. . . [E]ven if a judgment as to the abstract desirability of gradualism could be supported by evidence, it is outside the province of this Court to balance the merely desirable against the adjudicated constitutional rights of appellants. The Constitution has prescribed the educational policy applicable to the issue tendered in this case, and this Court has no power, under the guise of a "gradual" decree, to select another.

We submit that there are various necessary administrative factors

---

zona, New Mexico, and Kansas, 23 J. Negro Ed. 291, 293 (1954); Culver, Racial Desegregation in Education in Indiana, 23 J. Negro Ed. 296, 300–302 (1954).

which would make "immediate" relief as of tomorrow physically impossible. These include such factors as need for redistricting and the redistribution of teachers and pupils. Under the circumstances of this case, the Court's mandate will probably come down in the middle or near the close of the 1954 school term, and the decrees of the courts of first instance could not be put into effect until September, 1955. Appellees would, therefore, have had from May 17, 1954, to September, 1955 to make necessary administrative changes.

II

Answering Question 5: If This Court Should Decide to Permit an "Effective Gradual Adjustment" from Segregated School Systems to Systems Not Based on Color Distinctions, It Should Not Formulate Detailed Decrees but Should Remand These Cases to the Courts of First Instance with Specific Directions to Complete Desegregation by a Day Certain.

In answering Question 5, we are required to assume that this court "will exercise its equity powers to permit an effective gradual adjustment to be brought about from existing segregated systems to a system not based on color distinctions" thereby refusing to hold that appellants were entitled to decrees providing that, "within the limits set by normal geographic school districting, Negro children should forthwith be admitted to schools of their choice." While we feel most strongly that this Court will not subordinate appellants' constitutional rights to immediate relief to any plan for an "effective gradual adjustment," we must nevertheless assume the contrary for the purpose of answering Question 5.[12]

Question 5 assumes that there should be an "effective gradual adjustment" to a system of desegregated education. We have certain difficulties with this formulation. We have already demonstrated that there is no reason to believe that any form of gradualism will be more effective than forthwith compliance. If, however, this Court determines upon a gradual decree, we then urge that, as a minimum, certain safeguards must be embodied in that "gradual" decree in order to render it as nearly "effective" as any decree can be which continues the injury being suffered by these appellants as a consequence of the unconstitutional practice here complained of.

Appellants assume that "the great variety of local conditions," to which

---

[12]See question 5, page 158 in this volume.—ED.

the Court referred in its May 17th opinion, embraces only such educationally relevant factors as variations in administrative organization, physical facilities, school population and pupil redistribution, and does not include such judicially non-cognizable factors as need for community preparation . . . and threats of racial hostility and violence. . . .

Further we assume that the word "effective" might be so construed that a plan contemplating desegregation after the lapse of many years could be called an "effective gradual adjustment." For, whenever the change is in fact made, it results in a desegregated system. We do not understand that this type of adjustment would be "effective" within the meaning of Question 5 nor do we undertake to answer it in this framework. Rather, we assume that under any circumstances, the question encompasses due consideration for the constitutional rights of each of these appellants and those presently in the class they represent to be free from enforced racial segregation in public education.

Ordinarily, the problem—the elimination of race as the criterion of admission to public schools—by its very nature would require only general dispositive directions by this Court. Even if the Court decides that the adjustment to nonsegregated systems is to be gradual, no elaborate decree structure is essential at this stage of the proceedings. In neither event would appellants now ask this Court, or any other court, to direct or supervise the details of operation of the local school systems. In either event, we would seek effective provisions assuring their operation—forthwith in the one instance and eventually in the other—in conformity with the Constitution.

These considerations suggest appellants' answers to Question 5. Briefly stated, this Court should not formulate detailed decrees in these cases. It should not appoint a special master to hear evidence with a view to recommending specific terms for such decrees. It should remand these cases to the courts of first instance with directions to frame decrees incorporating certain provisions, hereinafter discussed, that appellants believe are calculated to make them as nearly effective as any gradual desegregation decree can be. The courts of first instance need only follow normal procedures in arriving at such additional provisions for such decrees as circumstances may warrant.

DECLARATORY PROVISIONS

This Court should reiterate in the clearest possible language that segregation in public education is a denial of the equal protection of the laws. It should order that the decrees include a recital that constitutional and statutory provisions, and administrative and judicial pronouncements, requiring or sanctioning segregated education afford no basis for the continued maintenance of segregation in public schools.

The important legal consequence of such declaratory provisions would be to obviate the real or imagined dilemma of some school officials who contend that, pending the issuance of injunctions against the continuation of segregated education in their own systems, they are entitled or even obliged to carry out state policies the invalidity of which this Court has already declared. . . .

TIME PROVISIONS

We do not know what considerations may be presented by appellees to warrant gradualism. But whatever these considerations may be, appellants submit that any school plan embracing gradualism must safeguard against the gradual adjustment becoming an interminable one. Therefore, appellants respectfully urge that this Court's opinion and mandate also contain specific directions that any decree to be entered by a district court shall specify (1) that the process of desegregation be commenced immediately, (2) that appellees be required to file periodic reports to the courts of first instance, and (3) an outer time limit by which desegregation must be completed.

Even cases involving gradual decrees have required some amount of immediate compliance by the party under an obligation to remedy his wrongs to the extent physically possible.[13]. . .

Whatever the reasons for gradualism, there is no reason to believe that the process of transition would be more effective if further extended. Certainly, to indulge school authorities until September 1, 1956, to achieve desegregation would be generous in the extreme. Therefore we submit that if the Court decides to grant further time, then the Court should direct that all decrees specify September, 1956, as the outside date by which desegregation must be accomplished. This would afford more than a year in excess of the time necessary for administrative changes, to review and modify decisions in the light of lessons learned as these decisions are put into effect.

We submit that the decrees should contain no provision for extension of the fixed limit, whatever date may be fixed. Such a provision would be merely an invitation to procrastinate.[14]

---

[13]See Wisconsin v. Illinois, 281 U. S. 179; Arizona Copper Co. v. Gillespie, 230 U. S. 46; Georgia v. Tennessee Copper Co., 206 U. S. 230; Westinghouse Air Brake Co. v. Great Northern Ry. Co., 86 Fed. 132 (C. C. S. D. N. Y. 1898).

[14]Ashmore, The Negro and the Schools 70–71 (1954); Culver, Racial Desegregation in Education in Indiana, 23 J. Negro Ed. 296–302 (1954).

We further urge this Court to make it plain that the time for completion of the desegregation program will not depend upon the success or failure of interim activities. The decrees in the instant cases should accordingly provide that in the event the school authorities should for any reason fail to comply with the time limitation of the decree, Negro children should then be immediately admitted to the schools to which they apply.[15] . . .

Conclusion

Much of the opposition to forthwith desegregation does not truly rest on any theory that it is better to accomplish it gradually. In considerable part, if indeed not in the main, such opposition stems from a desire that desegregation not be undertaken at all. In consideration of the type of relief to be granted in any case, due consideration must be given to the character of the right to be protected. Appellants here seek effective protection for adjudicated constitutional rights which are personal and present. Consideration of a plea for delay in enforcement of such rights must be preceded by a showing of clear legal precedent therefor and some public necessity of a gravity never as yet demonstrated.

[15]See United States v. American Tobacco Co., 221 U. S. 106, where this Court directed the allowance of a period of six months, with leave to grant an additional sixty days if necessary, for activities dissolving an illegal monopoly and recreating out of its components a new situation in harmony with the law, but further directed that if within this period a legally harmonious condition was not brought about, the lower court should give effect to the requirements of the Sherman Act [the first federal effort to deal with the growing problem of business monopolies (1890)].

Appellees' Brief
1954

. . . We think that a decree providing that, within the limits set by normal geographic school districting, Negro children should forthwith be admitted to schools of their choice, does not necessarily follow the opinion of May 17. On the other hand, we believe that this Court, in the exercise of its equity powers, may permit an effective gradual adjudgment from existing segregated systems to a system not based on color distinctions. The

Philip B. Kurland and Gerhard Casper, eds., *Landmark Briefs and Arguments of the Supreme Court of the United States: Constitutional Law,* vol. 49A (Arlington, Va.: University Publications of America, 1975), 684–85, 687–89.

very fact that these questions are now being argued, some seven months after the decision that segregation violates constitutional rights, suggests that the power to postpone compliance does exist.

The decree must seek to reconcile the personal and present interest of the Negro citizen, whose constitutional rights have been violated, with the public interest in safeguarding the integrity of the school system. To illustrate, we call attention specifically to a statement contained in the separate brief of the Board of Education of Topeka submitted prior to the December, 1953, arguments:

> If this Court should enter an order to abolish segregation in the public schools of Topeka, 'forthwith,' as suggested in Question 4(a), the Topeka Board would, of course, do its best to comply with the order. *We believe, however, that it would probably require that the regular classes be suspended, while the many administrative changes and adjustments are being made, and while the necessary transfers of and reassignment of students and teachers are being made.* Important decisions would have to be hurriedly made, without time for careful investigation of the facts nor for careful thought and reflection. Most decisions would have to be made on a temporary or an emergency basis. We believe the attendant confusion and interruption of the regular school program would be against the public interest, and would be damaging to the children, both negro and white alike. (pp. 405) (Italics supplied)

We think it cannot be disputed that a court of equity has power to avoid such a consequence.

The reports abound with authority for the proposition that it is the duty of a court of equity "to strike a proper balance between the needs of the plaintiff and the consequences of giving the desired relief."[1] . . .

This question compels our attention to the inherent limitations on the judicial power. We doubt that the Court contemplates the judicial development of a plan for the de-segregation of the schools of Kansas or any other state. If such action is contemplated, we doubt that it is legally or practically feasible. The Court may determine, as it has determined, that the segregated school system heretofore maintained in Topeka, Kansas, violates the Constitution of the United States. It may determine whether a gradual adjustment to a system not based on color distinctions is authorized. However, it cannot tell the Topeka Board of Education what non-segregated school system will be substituted for the one maintained, nor can it prescribe the course to be followed in effecting the substitution. These are determinations that must necessarily be made with reference to local conditions—conditions that were not germane to the question

---

[1] *Eccles v. Peoples Bank,* 333 U. S. 426, 431.

of whether segregation *per se* is unconstitutional and hence are not reflected by the record now before the Court. They are determinations that must be made by local officials who are familiar with local conditions and who are responsible for local educational policy and for the general administration of the school system. We urge that those officials be given the maximum latitude consistent with the rights of appellants. . . .

A review of the precedents would indicate that this Court, as a matter of policy, has heretofore refused to frame detailed decrees in cases involving segregation in education. In those cases where school facilities have been held unequal and where administrative action has been required to secure equality, the Court has not attempted to determine precise standards to be observed by the parties in order to finally dispose of the case. Rather, the Court has been content to remand the case to the lower court for further proceedings consistent with and in conformity with its opinion. . . .

We believe that the only order necessary in the present case, indeed, the only one justified by the circumstances, is one reversing the judgment of District Court, and remanding the cause to said court with directions to enter an appropriate decree. We suggest further that the District Court be directed to retain jurisdiction of the cause until such time as the maintenance of segregated schools by Appellee Board of Education is finally terminated. Implicit in such an order would be the power of the District Court upon appropriate motion by any of the parties to deal with special problems arising during the transition period.

Finally, we suggest that the decrees of both this Court and the District Court should provide for a minimum of judicial control. . . .

Wherever responsible state and local officials are proceeding in good faith to make the adjustments required by the Court's opinion of May 17, 1954, we suggest that their efforts be recognized and that they not be hedged by detailed judicial orders.

Appellants' Reply Brief
1954

Briefs Filed by Appellees and State Attorneys General Do Not Offer Any Affirmative Plan for Desegregation but Are Merely Restatements of Arguments in Favor of Interminable Continuation of Racial Segregation.

. . . [T]he briefs filed at this time, both by appellees and state attorneys general seems to be directed against ending racial segregation in our time, rather than toward desegregation within a reasonable time. First, these briefs do not in fact offer any affirmative plan or elements of such a plan for accomplishing the task of desegregation. Secondly, and equally significant, the main reasons now proffered in support of indefinite delay are identical with arguments previously advanced for denying relief on the merits.

This Court has decided that racial segregation is unconstitutional — that it is a practice, moreover, which has such effects on its victims that can only be described as abhorrent. Yet, in answering questions 4 and 5, propounded by the Court, the States do not even get around to what must, in the light of that decision, be the main problem underlying those questions: How can this practice be most expeditiously done away with? Reasons for delay, which would seem to occupy at best a subsidiary position, are the sole preoccupation of state counsel, and the affirmative problem gets virtually no attention.[1]

The brief of the Attorney General of Florida does contain a Point entitled "Specific Suggestions to the Court in Formulating a Decree."[2] But, the effect of the suggested plan[3] would be to subject the constitutional rights of Negro children to denial on the basis of such a variety of intangible factors that the plan itself cannot be seriously regarded as one for implementing the May 17th decision.

---

[1] It is true that Delaware and Kansas catalogue the progress they have made thus far in accomplishing integration. But both states plead for delay without offering any valid reasons therefor.

[2] Brief of the Attorney General of the State of Florida as *amicus curiae*, pp. 57–65. Hereinafter, citations to briefs of appellees and *amici curiae* will be abbreviated.

[3] Set out commencing at p. 61 of the Florida Brief.

Philip B. Kurland and Gerhard Casper, eds., *Landmark Briefs and Arguments of the Supreme Court of the United States: Constitutional Law*, vol. 49A (Arlington, Va.: University Publications of America, 1975), 707–23.

Each individual Negro child must, under the Florida plan, petition a court of the first instance for admission to an unsegregated school, after exhausting his administrative remedies. It is up to him to establish to that court's satisfaction that there exists no "reasonable grounds" for delay in his admission. "Reasonable grounds" include lack of a reasonable time to amend the state school laws, good faith efforts of the school board in promoting citizens' educational committees, administrative problems, and "evidence of . . . a strong degree of *sincere* opposition and sustained hostility" [emphasis supplied] giving the school board ground to believe that admission of the applicant would " . . . create emotional responses among the children which would seriously interfere with their education." In other words, the applicant's right is to be postponed until everything seems entirely propitious for granting it. It is submitted that this is not a plan for granting rights, but a plan for denying them just as long as can possibly be done without a direct overruling of the May 17th decision. . . .

The quality and thrust of the reasons now advanced for delay may best be evaluated by noting that (except for those that deal with purely administrative matters obviously requiring little time for solution) they are arguments which were advanced at an earlier stage in this litigation as grounds for denying relief on the merits, and now, under slightly altered guise, they walk again after their supposed laying to rest on May 17. Thus, the impossibility of procuring community acceptance of desegregation, urged earlier as a ground for decision on the merits,[4] now turns up as an argument for indefinite postponement[5] with no convincing reasons given for supposing that community attitudes will change within the segregated pattern.

The prediction that white parents will withdraw their children from public schools is repeated,[6] with the implied hope, no doubt, that at some remote date they will have attained a state of mind that will result in their leaving their children in school. "Racial tensions" are again predicted.[7] Negro teachers may lose their jobs.[8] Violence is warned of.[9] The people

[4]South Carolina Brief (1952) p. 27. Cf. *Id.* at p. 35; Virginia Brief (1952) pp. 24–25.

[5]Virginia Brief (1954) p. 13; Delaware Brief (1954) pp. 16, 25; Florida Brief (1954) p. 201 ff.; Texas Brief (1954) pp. 16–17; North Carolina Brief (1954) pp. 7–8.

[6]*Compare* Florida Brief (1954) pp. 26–27 and North Carolina Brief (1954) pp. 36–37 *with* Virginia Brief (1952) p. 30.

[7]*Compare* Florida Brief (1954) p. 95 *with* Virginia Brief (1952) p. 27.

[8]*Compare* Florida Brief (1954) pp. 31–32; North Carolina Brief (1954) pp. 24–25; and Texas Brief (1954) pp. 10–11, *with* Virginia Brief (1952) p. 31.

[9]*Compare* North Carolina Brief (1954) p. 37 and Florida Brief (1954) p. 25 *with* South Carolina Brief (1952) p. 27.

and the legislature will abolish the school system or decline to appropriate money for its support.[10]

All these are serious matters, but we have elsewhere shown solid reason for believing that those dire predictions, one and all, are unreliable. There is no reason for supposing that delay can minimize whatever unpleasant consequences might follow from the eradication of this great evil. Here, however, the point is that, where these arguments are resuscitated as grounds for delay, the inference is that their sponsors favor delay as long as present conditions prevail—that, in other words, they now want to delay desegregation just as long as the conditions exist which they formerly regarded as sufficient grounds for imposing segregation as a matter of legal right. The distinction is too fine to make such practical difference, either to the Negro child who is growing up or to this Court. . . .

Opinion Polls Are Immaterial to the Issues Herein and Do Not Afford Any Basis to Support an Argument That a Gradual Adjustment Would Be More Effective.

Several of the briefs filed herein refer to polls of public opinion in their respective States in support of arguments to postpone desegregation indefinitely.[11] These polls appear to have been made for the purpose of sampling opinions of various groups within the State as to whether they approved of the May 17th decision and whether they thought it could be enforced immediately without friction.

The information as to racial hostility obtained from these polls is indecisive of the issues before this Court. In *Buchanan* v. *Warley,* 245 U. S. 60, 80, this Court stated:

> That there exists a serious and difficult problem arising from a feeling of race hostility which the law is powerless to control, and to which it must give a measure of consideration, may be freely admitted. But its solution cannot be promoted by depriving citizens of their constitutional rights and privileges.

We believe the same answer should be given to any suggestion that the enforcement of constitutional rights be deferred to a time when it will have uniform public acceptance. . . .

Moreover, such polls are not a valid index of how the individuals questioned will in fact act in the event of desegregation. Modern psychologi-

[10]*Compare* North Carolina Brief (1954) p. 36; Virginia Brief (1954) p. 15; and Arkansas Brief (1954) pp. 7–8 *with* South Carolina Brief (1952) p. 27.

[11]Texas Brief pp. 16–17; Virginia Brief pp. 13–14; North Carolina Brief pp. 7–9; Florida Brief pp. 23–24, 105 ff; Delaware Brief p. 12.

cal research shows that, especially in the case of broad public issues, many persons simply "do not follow through even on actions which they say they personally will take in support of an opinion."[12] . . .

Finally, there is nothing to indicate that an extended delay in ordering the elimination of all segregation will improve public attitudes or eliminate the objections presently interposed. Clearly the polls are irrelevant and should be so treated by this Court.

The Wide Applicability of the Decision in These Cases Should Not Affect the Relief to Which Appellants Are Entitled.

Effort is made throughout the briefs for appellees and the several attorneys general to balance the personal and present rights here involved against the large number of children of both races now attending public school on a segregated basis. This argument is made for a twofold purpose: to escape the uniformity of decisions of this Court on the personal character of the rights involved and, secondly, to destroy the present character of the right involved.

Of course, the decision of this Court in the instant cases will have wide effect involving public school systems of many states and many public school children. The mere fact of numbers involved is not sufficient to delay enforcement of rights of the type here involved.[13]

On the face of it, their position is both ill-taken and self-defeating. That it is ill-taken becomes clear when the suggestion itself is clearly stated; obviously, there is nothing in mere numerousness as such which has any tendency whatever to create or destroy rights to efficacious legal relief. Behind every numeral is a Negro child, suffering the effects spoken of

---

[12]Buchanan, Krugman and Van Wagenen, An International Police Force and Public Opinion 13 (1954). For other studies dealing with the discrepancy between verbal statements and actions, see Link and Freiberg, "The Problem of Validity vs. Reliability in Public Opinion Polls", 6 Public Opinion Quarterly 87–98, esp. 91–92 (1942); Jenkins and Corbin, "Dependability of Psychological Brand Barometers II, The Problem of Validity", 22 Journal of Applied Psychology 252–260 (1938); Hyman, "Do They Tell the Truth?", 8 Public Opinion Quarterly 557–559 (1944); Social Science Research Council, Committee on Analysis of Pre-Election Polls and Forecasts 302–303 (1949); La Piere, "Attitudes vs. Actions," 13 Social Forces 230–237 (1934); Doob, Public Opinion and Propaganda 151 (1948); Hartley and Hartley, Fundamentals of Social Psychology 657 (1952). See also *Irvin* v. *State,* 66 So. 2d 288, 290–292, *cert. denied* 346 U. S. 927, *reh. denied* 347 U. S. 914.

[13]We put to one side as obviously immaterial the mere technical character of these suits as class actions as under Rule 23(a)(3). Obviously, the mere joinder of plaintiffs in a spurious class suit for reasons of convenience cannot have any effect on the nature of the rights asserted or on the availability of normal relief remedy. Whether a suit is or is not a class action tells us little, in this field of law, as to the magnitude of the interests involved; *Sweatt* v. *Painter* was an individual mandamus suit [one seeking a court-ordered command that a specific action be taken], but the effect of that decision spread throughout the segregating states.

by the Court on May 17. It is a manifest inconsequence to say that the rights or remedial needs of each child are diminished merely because others are in the same position. That this argument is self-defeating emerges when it is considered that its tendency is simply to establish that we have to do with an evil affecting a great many people; presumably, the abolition of a widespread evil is even more urgent than dealing with isolated cases of wrongdoing.

This Court has consistently treated the personal rights of litigants on a personal basis. Every leading case involving discrimination against Negroes has necessarily and demonstrably involved large numbers of people; yet this Court has given present relief on a personal basis to those who showed themselves entitled to it, without any hint of the possibility that the rights of citizenship are diminished because many people are being denied them. . . . All major constitutional cases involve large numbers of people. Yet there is not a hint, in words or in action, in any past case, to the effect that the wide applicability of a decision was considered material to the right to relief. It is unthinkable that this Court would apply any such doctrine to limit the enjoyment of constitutional rights in general; there is no reason for its making a special and anomalous exception of the case at bar. . . .

Average Differences in Student Groups Have No Relevance to the Individual Rights of Pupils: Individual Differences Can Be Handled Administratively Without Reference to Race.

. . . We have come too far not to realize that educability and absorption and adoption of cultural values has nothing to do with race. What is achieved educationally and culturally, we now know to be largely the result of opportunity and environment.[14] That the Negro is so disadvantaged educationally and culturally in the states where segregation is required is the strongest argument against its continuation for any period of time. Yet those who use this argument as a basis for interminable delay in the elimination of segregation in reality are seeking to utilize the product of their own wrongdoing as a justification for continued malfeasance.

Our public school systems have grown and improved as an American institution. And in every community it is obvious that children of all levels of culture, educability, and achievement must be accounted for within

---

[14]Klineberg, Race Differences: The Present Position of the Problem, 2 International Social Science Bulletin 460 (1950); Montague, Statement on Race, The Unesco Statement by Experts on Race Problems 14–15 (1951); Montague, Man's Most Dangerous Myth: The Fallacy of Race 286 (1952); Kirkpatrick, Philosophy of Education 399–433 (1951). See Klineberg, Race and Psychology, Unesco (1951); Allport, The Nature of Prejudice (1954); Comas, Racial Myths, Unesco (1951).

the same system. In some school systems the exceptional children are separated from the rest of the children. In others there are special classes for retarded children, for slow readers and for the physically handicapped. But these factors have no relation to race. These are administrative problems with respect to conduct of the public school.

In the past, large city school systems, North and South, have had the problem of absorbing children from rural areas where the public schools and cultural backgrounds were below city standards. On many occasions these migrations have been very sudden and in proportionately very large numbers. This problem has always been solved as an administrative detail. It has never been either insurmountable or has it been used as an excuse to force the rural children to attend sub-standard schools. Similarly, large cities have met without difficulty the influx of immigrants from foreign countries. . . .

Official Reactions in States Affected by the May 17th Decision Make It Plain that Delay Will Detract from Rather Than Contribute to the "Effectiveness" of the Transition to Desegregated Schools.

Events occurring in the states affected by the decision of May 17, 1954, do not support the suggestions of appellees and *amici curiae* that further (and limitless) postponement of relief to Negro children will assure an "effective" adjustment from segregated to non-segregated school systems. In terms of legislative, executive or administrative reaction, the southern and border states may now be grouped in three loose categories:

(1) Those which have not waited for further directions from the Court, but have undertaken desegregation in varied measure during the current school year. Typical of the states falling in this category are Delaware,[15] Kansas,[16] Missouri,[17] and West Virginia.[18] Although not a state, the District of Columbia would fall within this group.

(2) Those which have decided to await a decision on the question of

---

[15]Brief for Appellants in Nos. 1, 2 and 3 and for Respondents in No. 5 on Further Reargument, pp. 4–7; Brief for Petitioners on the Mandate in No. 5, pp. 10–12.

[16]Brief for Appellants in Nos. 1, 2 and 3 and for Respondents in No. 5 on Further Reargument, pp. 3–4; Supplemental Brief for the State of Kansas on Questions 4 and 5 Propounded by the Court, pp. 13–22; Supplemental Brief for the Board of Education, Topeka, Kansas on Questions 4 and 5 Propounded by the Court, pp. 2–4.

[17]Southern School News, September 3, 1954, p. 9, c. 2–5; *Id.,* October 1, 1954, p. 10, c. 1–5; *Id.,* January 6, 1955, p. 11, c. 1; *Id.,* February 3, 1955, p. 15, c. 1–5.

[18]Southern School News, October 1, p. 14, c. 1, 5; *Id.,* January 6, 1955, p. 2, c. 4–5.

relief but have indicated an intention to obey the Court's directions. Kentucky,[19] Oklahoma,[20] and Tennessee[21] are among the states in this category.

(3) Those which have indicated an intention to circumvent the decision of this Court or interminably delay the enjoyment by Negro children of their constitutionally protected rights not to be segregated in public schools. Included in this category are states like South Carolina[22] and Mississippi,[23] which have enacted legislation designed to nullify any decision of this Court in these cases, and states like Virginia[24] and Florida,[25] where either the governors or special legislative committees studying the problem have recommended that "every legal means" be used to preserve segregated school systems.[26]

Against this background of state reaction to the decision of May 17, 1954, it is clear that postponement of relief will serve no purpose. The states in the first category have already begun to implement this Court's decision and any delay as to them may imperil the progress already made.[27] The states in the second category have indicated a willingness to do whatever this Court directs and there is certainly no reason for delay as to them. The probable effect of delay, as to states in the third category, must be evaluated in the light of their declared intentions; we are justified in assuming that it would have no affirmative effect, but would merely provide additional time to devise and put into practice schemes expressly designed to thwart this Court's decision.

[19]Southern School News, September 3, 1954, p. 7, c. 3; *Id.*, November 4, 1954, p. 16, c. 1; *Id.*, December 1, 1954, p. 9, c. 1, 3.
[20]Southern School News, February 3, 1955, p. 10, c. 1–2; *Id.*, March 3, 1955, p. 16, c. 1; The New York Times, April 6, 1955, p. 20, c. 5.
[21]Southern School News, October 1, 1954, p. 11, c. 1; *Id.*, December 1, 1954, p. 12, c. 4; New York Post, March 16, 1955, p. 58, c. 4.
[22]Southern School News, September 3, 1954, p. 12, c. 1–2; *Id.*, February 3, 1955, p. 3, c. 2–4; *Id.*, March 3, 1955, p. 14, c. 1–3.
[23]Southern School News, September 3, 1954, p. 8, c. 3; *Id.*, October 1, 1954, p. 9, c. 4–5; *Id.*, November 4, 1954, p. 11, c. 4–5; *Id.*, January 6, 1955, p. 10, c. 1–2; The New York Times, April 6, 1955, p. 20, c. 5.
[24]Southern School News, February 3, 1955, p. 10, c. 4.
[25]Southern School News, January 6, 1955, p. 6, c. 2.
[26]Indeed, Governor Marvin B. Griffin of Georgia has asserted: "However, if this court is so unrealistic as to attempt to enforce this unthinkable evil upon us, I serve notice now that we shall resist it with all the resources at our disposal and we shall never submit to the proposition of mixing the races in the classrooms of our schools."
[27]See, *e.g. Steiner* v. *Simmons*, 111 A. 2d 574 (Del. 1955), rev'g. 108 A. 2d 173 (Del. 1954). There the Supreme Court reversed a chancery court determination that forthwith desegregation was proper under the decision of this Court of May 17, 1954.

Conclusion

. . . [I]t should be borne in mind that the very magnitude of these problems exists because of the assumption, tacitly indulged up to now, that the Constitution is not to be applied in its full force and scope to all sections of this country alike, but rather that its guarantees are to be enjoyed, on one part of our nation, only as molded and modified by the desire and customs of the dominant component of the sectional population. Such a view, however expressed, ignores the minimum requirement for a truly national constitution. It ignores also a vast part of the reality of the sectional interest involved, for that interest must be composed of the legitimate aspirations of Negroes as well as whites. It certainly ignores the repercussions which any reluctance to forthrightly enforce appellants' rights would have on this nation's international relations. Every day of delay means that this country is failing to develop its full strength.

The time has come to end the division of one nation into those sections where the Constitution is and those where it is not fully respected. Only by forthright action can the country set on the road to a uniform amenability to its Constitution. Finally, the right asserted by these appellants is not the only one at stake. The fate of other great constitutional freedoms, whether secured by the Fourteenth Amendment or by other provisions, is inevitably bound up in the resolution to be made in these cases. For delay in enforcement of these rights invites the insidious prospect that a moratorium may equally be placed on the enjoyment of other constitutional rights.

CHIEF JUSTICE EARL WARREN

Ruling on Relief
May 31, 1955

These cases were decided on May 17, 1954. The opinions of that date,[1] declaring the fundamental principle that racial discrimination in public education is unconstitutional, are incorporated herein by reference. All provisions of federal, state, or local law requiring or permitting such dis-

[1]347 U.S. 483; 347 U.S. 497.

Brown et al. v. Board of Education of Topeka et al., 349 US 294 (1955).

crimination must yield to this principle. There remains for consideration the manner in which relief is to be accorded.

Because these cases arose under different local conditions and their disposition will involve a variety of local problems, we requested further argument on the question of relief.[2] In view of the nationwide importance of the decision, we invited the Attorney General of the United States and the Attorneys General of all states requiring or permitting racial discrimination in public education to present their views on that question. The parties, the United States, and the States of Florida, North Carolina, Arkansas, Oklahoma, Maryland, and Texas filed briefs and participated in the oral argument.

These presentations were informative and helpful to the Court in its consideration of the complexities arising from the transition to a system of public education freed of racial discrimination. The presentations also demonstrated that substantial steps to eliminate racial discrimination in public schools have already been taken, not only in some of the communities in which these cases arose, but in some of the states appearing as *amici curiae,* and in other states as well. Substantial progress has been made in the District of Columbia and in the communities in Kansas and Delaware involved in this litigation. The defendants in the cases coming to us from South Carolina and Virginia are awaiting the decision of this Court concerning relief.

Full implementation of these constitutional principles may require solution of varied local school problems. School authorities have the primary responsibility for elucidating, assessing, and solving these problems; courts will have to consider whether the action of school authorities constitutes good faith implementation of the governing constitutional principles. Because of their proximity to local conditions and the possible need for further hearings, the courts which

---

[2]Further argument was requested on the following questions, 347 U.S. 483, 495–496, n. 13, previously propounded by the Court:

"4. Assuming it is decided that segregation in public schools violates the Fourteenth Amendment

"*(a)* would a decree necessarily follow providing that, within the limits set by normal geographic school districting, Negro children should forthwith be admitted to schools of their choice, or

"*(b)* may this Court, in the exercise of its equity powers, permit an effective gradual adjustment to be brought about from existing segregated systems to a system not based on color distinctions?

"5. On the assumption on which questions 4*(a)* and *(b)* are based, and assuming further that this Court will exercise its equity powers to the end described in question 4 *(b)*,

"*(a)* should this Court formulate detailed decrees in these cases;

"*(b)* if so, what specific issues should the decrees reach;

These photographs showcase the centrality of the struggle against Jim Crow restrictions in intrastate (as well as interstate) public conveyances, notably buses. On April 25, 1956, the Dallas Transit Company complied with the Supreme Court ban on racial segregation in public intrastate carriers by integrating its buses. Both photos celebrate the triumph over Jim Crow: represented here especially by Jim Crow's omnipresent language, signs, and symbols. The top photo revels in the moment of that first ride in the formerly all-white seating area under a now anachronistic Jim Crow sign. The bottom photo freezes both the moment of that sign's demise and the fall of the system represented by that very sign.

originally heard these cases can best perform this judicial appraisal. Accordingly, we believe it appropriate to remand the cases to those courts.[3]

In fashioning and effectuating the decrees, the courts will be guided by equitable principles. Traditionally, equity has been characterized by a practical flexibility in shaping its remedies[4] and by a facility for adjusting and reconciling public and private needs.[5] These cases call for the exercise of these traditional attributes of equity power. At stake is the personal interest of the plaintiffs in admission to public schools as soon as practicable on a nondiscriminatory basis. To effectuate this interest may call for elimination of a variety of obstacles in making the transition to school systems operated in accordance with the constitutional principles set forth in our May 17, 1954, decision. Courts of equity may properly take into account the public interest in the elimination of such obstacles in a systematic and effective manner. But it should go without saying that the vitality of these constitutional principles cannot be allowed to yield simply because of disagreement with them.

While giving weight to these public and private considerations, the courts will require that the defendants make a prompt and reasonable start toward full compliance with our May 17, 1954, ruling. Once such a start has been made, the courts may find that additional time is necessary to carry out the ruling in an effective manner. The burden rests upon the defendants to establish that such time is necessary in the public interest and is consistent with good faith compliance at the earliest practicable date. To that end, the courts may consider problems related to administration, arising from the physical condition of the school plant, the school transportation system, personnel, revision of school districts and attendance areas into compact units to achieve a system of determining admission to the public schools on a nonracial basis, and revision of local laws and regulations which may be necessary in solving the foregoing problems. They will also consider the adequacy of any plans the defendants may propose to meet these problems and to effectuate a transition

---

"*(c)* should this Court appoint a special master to hear evidence with a view to recommending specific terms for such decrees;

"*(d)* should this Court remand to the courts of first instance with directions to frame decrees in these cases, and if so what general directions should the decrees of this Court include and what procedures should the courts of first instance follow in arriving at the specific terms of more detailed decrees?"

[3]The cases coming to us from Kansas, South Carolina, and Virginia were originally heard by three-judge District Courts convened under 28 U.S.C. §§ 2281 and 2284. These cases will accordingly be remanded to those three-judge courts. See *Briggs* v. *Elliott,* 342 U.S. 350.

[4]See *Alexander* v. *Hillman,* 296 U.S. 222, 239.

[5]See *Hecht Co.* v. *Bowles,* 321 U.S. 321, 329–330.

to a racially nondiscriminatory school system. During this period of transition, the courts will retain jurisdiction of these cases.

The judgments below . . . are accordingly reversed and the cases are remanded to the District Courts to take such proceedings and enter such orders and decrees consistent with this opinion as are necessary and proper to admit to public schools on a racially nondiscriminatory basis with all deliberate speed the parties to these cases. . . .

5

Popular Response to *Brown*

NEWSPAPER EDITORIALS

The following documents provide a national cross-section of editorial opinion in response to the 1954 *Brown* decision. While clearly reflecting the views of the individual editors and editorial boards that crafted them, these editorials necessarily sought in part to represent the moods and reactions of their constituencies. It is useful to consider these documents as rhetorical efforts aimed at communicating general understandings within specific communities: cultural and journalistic communities on one hand and national, regional, and local communities on the other. In this case, editorial authority derived in part from how well these writers both responded to and represented the prevailing climate of opinion.

These editorials were all cited in the *New York Times,* the acknowledged newspaper of record and thus the most powerful editorial voice in the nation. Various other journalistic worlds are represented. The Atlanta *Daily World,* Pittsburgh *Courier,* and Chicago *Defender* are black newspapers. The Washington *Post and Times Herald,* Atlanta *Constitution,* Jackson (Mississippi) *Daily News,* and Arkansas *Gazette* are southern white newspapers. In addition to the New York *Times,* the Boston *Herald,* Los Angeles *Times,* and Chicago *Sun-Times* are white newspapers outside the South. Finally, white student editorial opinion is represented by newspapers from the University of Virginia and the University of Mississippi.

Do you detect any recognizable patterns—for instance, regional, big city versus small city, black versus white—in the editorials? Do you see a central concern, or primary argument, running through them? Are there salient similarities and differences? What kind of impact do you think these editorials had at the time? How do you assess their viability as historical evidence?

"All God's Chillun"

May 18, 1954

The Supreme Court took a long and careful time to arrive at the unanimous decision read yesterday by Chief Justice Warren that "segregation of children in the public schools solely on the basis of race, even though the physical facilities and other 'tangible' factors may be equal, deprives the children of the minority group of equal educational opportunities." But the decision reached was inevitable in the year 1954 regardless of what may have been the case in 1868, when the Fourteenth Amendment was adopted, or in 1896, when the "separate but equal" doctrine was laid down in the case of *Plessy v. Ferguson.*

In the cases under consideration the facilities offered to Negro children appeared to be equal, or were to be made equal, "with respect to buildings, curricula, qualifications and salaries of teachers and other 'tangible' factors," to those available to white children. The question, therefore, was more fundamental than in any previous case. It was whether Negro children segregated solely on the basis of race, even though offered equal facilities, were thereby deprived of equal educational opportunities. The court holds that such segregation does have "a detrimental effect upon the colored children," that it had "a tendency to retard [their] educational and mental development . . . and to deprive them of some of the benefits they would receive in a racially integrated school system."

The court, speaking through Chief Justice Warren, therefore concludes that "separate educational facilities are inherently unequal," that the plaintiffs and others similarly situated "are by reason of the segregation complained of deprived of the equal protection of the laws guaranteed by the Fourteenth Amendment." The due process clause is not involved. It is not needed.

What the court is saying, in its formal but not complicated style, is a part of what Eugene O'Neill said in a play called *All God's Chillun Got Wings.* It is true, of course, that the court is not talking of that sort of "equality" which produces interracial marriages. It is not talking of a social system at all. It is talking of a system of human rights which is foreshadowed in the second paragraph of the Declaration of Independence, which stated "that all men are created equal." Mr. Jefferson and the others who were responsible for the Declaration did not intend to say that all men are equally intelligent, equally good or equal in height or weight.

They meant to say that men were, and ought to be, equal before the law. If men are equal, children are equal, too. There is an even greater necessity in the case of children, whose opportunities to advance themselves and to be useful to the community may be lost if they do not have the right to be educated.

No one can deny that the mingling of the races in the schools of the seventeen states which have required segregation and the three states which have permitted it will create problems. The folkways in southern communities will have to be adapted to new conditions if white and Negro children, together with white and Negro teachers, are to enjoy not only equal facilities but the same facilities in the same schools. The Constitution and the Bill of Rights are at times hard masters. The court has recognized these difficulties by withholding a decree and by inviting "the full assistance of the parties in formulating decrees." The cases are therefore restored to the docket and the Attorney General of the United States and the Attorneys General of the states requiring or permitting segregation in public education will be permitted to appear before the court next fall. There will be some delay before orders issue, and it may be that petitions for rehearing and modification will take up a good deal of time. These matters cannot be hurried.

A constitutional principle inherent in the Declaration of Independence and never entirely forgotten, even in the days of human slavery, has, however, been restated. This nation is often criticized for its treatment of racial minorities, and particularly of the Negro. There have been grounds for this criticism. Little by little, however, in the folk customs and in such decisions as the one rendered yesterday, we move toward a more perfect democracy. When some hostile propagandist rises in Moscow or Peiping to accuse us of being a class society we can if we wish recite the courageous words of yesterday's opinion. The highest court in the land, the guardian of our national conscience, has reaffirmed its faith—and the undying American faith—in the equality of all men and all children before the law.

DAILY WORLD (ATLANTA)

The Decision Of A Century

May 18, 1954

... This case has attracted world attention; its import will be of great significance in these trying times when democracy itself is struggling to envision a free world. It will strengthen the position of our nation in carrying out the imposed duties of world leadership.

Coming at this particular time, the decision serves as a boost to the spirit of Democracy, it accelerates the faith of intense devoutness in minorities, who have long believed in and trusted the courts.

However, it has added significance to the citizens of Georgia who are now confronted with a proposed state constitutional amendment to turn the schools from public to private hands in the event the court did just what it has done. We predict now the defeat of this amendment.

COURIER (PITTSBURGH)

Will Stun Communists

May 18, 1954

The conscience of America has spoken through its constitutional voice.... This clarion announcement will also stun and silence America's Communist traducers behind the Iron Curtain. It will effectively impress upon millions of colored people in Asia and Africa the fact that idealism and social morality can and do prevail in the United States, regardless of race, creed or color.

As reprinted in "Editorial Excerpts from the Nation's Press on Segregation Ruling," *New York Times,* May 18, 1954.

DEFENDER (CHICAGO)

End of Dual Society

May 18, 1954

Neither the atom bomb nor the hydrogen bomb will ever be as meaningful to our democracy as the unanimous declaration of the Supreme Court that racial segregation violates the spirit and the letter of our Constitution. This means the beginning of the end of the dual society in American life and the . . . segregation which supported it.

As reprinted in "Editorial Excerpts from the Nation's Press on Segregation Ruling," *New York Times*, May 18, 1954.

POST AND TIMES HERALD (WASHINGTON, D.C.)

Emancipation

May 18, 1954

The Supreme Court's resolution yesterday of the school segregation cases affords all Americans an occasion for pride and gratification. The decision will prove, we are sure—whatever transient difficulties it may create and whatever irritations it may arouse—a profoundly healthy and healing one. It will serve—and speedily—to close an ancient wound too long allowed to fester. It will bring to an end a painful disparity between American principles and American practices. It will help to refurbish American prestige in a world which looks to this land for moral inspiration and restore the faith of Americans themselves in their own great values and traditions.

The Supreme Court Has Given Us Time

May 18, 1954

. . . The court decision does not mean that Negro and white children will go to school together this fall. The court itself provides for a "cooling off" period. Not until next autumn will it even begin to hear arguments from the attorneys general of the 17 states involved on how to implement the ruling.

Meanwhile, it is no time for hasty or ill-considered actions. It is no time to indulge demagogues on either side nor to listen to those who always are ready to incite violence and hate.

It is a time for Georgia to think clearly. Our best minds must be put to work, not to destroy, but to seek out constructive conclusions.

DAILY NEWS (JACKSON MISS.)

Bloodstains On White Marble Steps

May 18, 1954

. . . Human blood may stain Southern soil in many places because of this decision but the dark red stains of that blood will be on the marble steps of the United States Supreme Court building.

White and Negro children in the same schools will lead to miscegenation. Miscegenation leads to mixed marriages and mixed marriages lead to mongrelization of the human race.

The Pattern of the Future
May 18, 1954

. . . "In many ways this is the end product of a process that began many years ago and began, as such things must, at the very top of the educational structure. But where a half-century ago Booker T. Washington and a handful of white Southern college presidents represented the outside limits of free and frank discussion of the mutual problems inherent in educating the two races, the level of discussion has now descended to the cross-roads hamlet and the participants are not learned educators but the parents of Negro children and the white laymen who sit on a local school board.

"It is here that the South will have to determine the future of its educational system. Wise leadership at the upper levels can help, and emotional excursions by the leaders of either race can do great harm. But in the end the new patterns will have to be hammered out across the table in thousands of scattered school districts, and they will have to be shaped to accommodate not only the needs but the prejudices of whites and Negroes to whom these problems are not abstractions but the essence of their daily lives."

Equality Redefined
May 18, 1954

The Supreme Court's history-making decision against racial segregation in the public schools proves more than anything else that the Constitution is still a live and growing document.

. . . The segregation ruling is frankly expedient. It recognizes the growing national feeling that the separation of Negro (or other minority) children from the majority race at school age is an abuse of the democratic process and the democratic principle. But it is also the culmination of a series of judicial opinions which circumspectly prepared the way for change.

TIMES (LOS ANGELES)
The Segregation Decision
May 19, 1954

... We may be sure that the present decision on segregation is not going to lead to civil war, but we may be almost as certain that it will provoke a social and political revolution. Yet it is hard at this point to see how the court could have come to any other conclusion.

Enforcement of the decision looms as a tremendous problem for the court. Apparently it will require pressure on, or coercion of, virtually the whole white populations of the States where segregation has been law and custom.

... The confusion of change in some areas could work to the disadvantage of at least one school generation, Negro and white, unless mutual restraint and understanding are joined in resolving the issue.

CAVALIER DAILY (UNIVERSITY OF VIRGINIA)
"Violates" Way of Life
May 18, 1954

It is too early to tell what effect the Supreme Court decision to abolish segregated schools will have on the South. . . . Although it is hard from a strict legal point of view to justify any action contrary to law, we feel that

As reprinted in "Editorial Excerpts from the Nation's Press on Segregation Ruling," *New York Times,* May 18, 1954.

the people of the South are justified in their bitterness concerning this decision. To many people this decision is contrary to a way of life and violates the way in which they have thought since 1619.

MISSISSIPPIAN (UNIVERSITY OF MISSISSIPPI)

Adjustment Held Difficult

May 18, 1954

We realize that the decision was a difficult one at which to arrive, and we hope that the Supreme Court fully realizes that the adjustment will be much more difficult than was its decision. We know that the student body of the University of Mississippi has long been aware of the problem and its complexities, and has accepted the fact that Negroes will probably some day be admitted into the university. Though the majority of the students do not want to attend school with Negroes, we feel that the students will adapt themselves to it.

LETTERS TO EDITORS

The range of individual responses to *Brown* ran the gamut. To some it was the second Emancipation Proclamation. To others, it was the most serious challenge to white supremacy since blacks participated in Reconstruction — or, as this camp grossly misrepresented that historical moment, the dreadful era of black (and Yankee) domination over a prostrate white South.

Included here are two provocative and iconoclastic letters to the editor from southerners: anthropologist, folklorist, and writer Zora Neale Hurston (1891–1960) and writer and social critic Lillian Smith (1897–1966). The latter was white, the former black. Smith was an enthusiastic supporter of integration, and in 1955 she wrote a ringing endorsement of *Brown* in her book *Now Is the Time.* Hurston vigorously opposed

As reprinted in "Editorial Excerpts from the Nation's Press on Segregation Ruling," *New York Times,* May 18, 1954.

Brown and, as demonstrated here, ardently supported racially separate institutions, cultures, and social lives. Both Hurston and Smith at the time voiced a view diametrically opposed to the dominant opinion of black and white southerners, respectively.

What are the common threads knitting these letters together? What are the salient differences? How do you explain these similarities and differences? Do you find any or all of Hurston's arguments persuasive? Why or why not? In what direction and to what effect does Smith expand the definition of equality of educational opportunity? Finally, how does the cold war figure into these letters?

TIMES (NEW YORK)

Ruling on Schools Hailed
May 31, 1954

To the Editor of the New York Times:

I have read again the recent decision of the Supreme Court. It bears rereading. For it is a great historic document—not only because its timing turns it into the most powerful political instrument against communism that the United States has as yet devised but because of its profound meaning for children.

It is every child's Magna Charta. All are protected by the magnificent statement which declares that no artificial barriers, such as laws, can be set up in our land against a child's right to learn and to relate himself to his world.

There are perhaps five million children in the United States who are colored. There are close to five million other children who will be directly affected by this decision. I am not speaking of the majority of white children, many of whom have undoubtedly been injured spiritually by the philosophy and practice of segregation. I am speaking of disabled children, who are "different," not because of color but because of blindness, deafness; because they are crippled, have cerebral palsy, or speech defects, or epilepsy; or are what we call "retarded." These children we have also segregated.

There are more than forty states with laws forbidding a child with epilepsy to attend public school—even though most children's convulsions can now be controlled by modern drugs. Many blind children are segregated in schools from sighted children; our deaf, from the hearing. Many cerebral palsy children are kept out of school not because they are unable to attend but because there are teachers who do not want to teach them. And yet a basic principle of rehabilitation is that acceptance by others and a natural relationship with his human world are necessary for the disabled child if he is to make a good life for himself.

All these children, some with real disabilities, others with the artificial disability of color, are affected by this great decision.

Then why are a few politicians protesting so angrily? Perhaps because they feel they will now be handicapped if the old crutch of "race" is snatched away from them.

It is true that this decision may, in certain parts of the country, shackle a few politicians. But it frees so many of our children. I for one am glad. And I believe millions of other Southerners are glad also, and will accept wholeheartedly the challenge of making a harmonious, tactful changeover from one kind of school to another. It will be an ordeal only if our attitude makes it one. There are creative, practical ways of bringing about this change, whether we live North or South. And in the doing of it we adults may grow, too, in wisdom and gentleness.

—Lillian Smith
Clayton, Ga.

SENTINEL (ORLANDO)

Court Order Can't Make Races Mix

August 11, 1955

Editor: I promised God and some other responsible characters, including a bench of bishops, that I was not going to part my lips concerning the U. S. Supreme Court decision on ending segregation in the public schools of the South. But since a lot of time has passed and no one

Zora Neale Hurston, "Court Order Can't Make Races Mix," *Orlando Sentinel, The Public Thought,* Aug. 11, 1955.

seems to touch on what to me appears to be the most important point in the hassle, I break my silence just this once. Consider me as just thinking out loud.

The whole matter revolves around the self-respect of my people. How much satisfaction can I get from a court order for somebody to associate with me who does not wish me near them? The American Indian has never been spoken of as a minority and chiefly because there is no whine in the Indian. Certainly he fought, and valiantly for his lands, and rightfully so, but it is inconceivable of an Indian to seek forcible association with anyone. His well known pride and self-respect would save him from that. I take the Indian position.

Now a great clamor will arise in certain quarters that I seek to deny the Negro children of the South their rights, and therefore I am one of those "handkerchief-head niggers" who bow low before the white man and sell out my own people out of cowardice. However an analytical glance will show that that is not the case.

If there are not adequate Negro schools in Florida, and there is some residual, some inherent and unchangeable quality in white schools, impossible to duplicate anywhere else, then I am the first to insist that Negro children of Florida be allowed to share this boon. But if there are adequate Negro schools and prepared instructors and instructions, then there is nothing different except the presence of white people.

For this reason, I regard the ruling of the U. S. Supreme Court as insulting rather than honoring my race. Since the days of the never-to-be-sufficiently-deplored Reconstruction, there has been current the belief that there is no greater delight to Negroes than physical association with whites. The doctrine of the white mare. Those familiar with the habits of mules are aware that any mule, if not restrained, will automatically follow a white mare. Dishonest mule-traders made money out of this knowledge in the old days.

Lead a white mare along a country road and slyly open the gate and the mules in the lot would run out and follow this mare. This ruling being conceived and brought forth in a sly political medium with eyes on '56, and brought forth in the same spirit and for the same purpose, it is clear that they have taken the old notion to heart and acted upon it. It is a cunning opening of the barnyard gate with the white mare ambling past. We are expected to hasten pell-mell after her.

It is most astonishing that this should be tried just when the nation is exerting itself to shake off the evils of Communist penetration. It is to be recalled that Moscow, being made aware of this folk belief, made it the main plank in their campaign to win the American Negro from the 1920s on. It was the come-on stuff. Join the party and get yourself a white wife or husband. To supply the expected demand, the party had scraped up this-and-that off of park benches and skid rows and held them in stock for us. The highest types of Negroes were held to be just panting to get hold of one of these objects. Seeing how flat that program fell, it is astonishing that it would be so soon revived. Politics does indeed make strange bedfellows.

But the South had better beware in another direction. While it is being frantic over the segregation ruling, it had better keep its eyes open for more important things. One instance of Govt by fiat has been rammed down its throat. It is possible that the end of segregation is not here and never meant to be here at present, but the attention of the South directed on what was calculated to keep us busy while more ominous things were brought to pass. The stubborn South and the Midwest kept this nation from being dragged farther to the left than it was during the New Deal.

But what if it is contemplated to do away with the two-party system and arrive at Govt by administrative decree? No questions allowed and no information given out from the administrative dept? We could get more rulings on the same subject and more far-reaching any day. It pays to weigh every saving and action, however trivial as indicating a trend.

In the ruling on segregation, the unsuspecting nation might have witnessed a trial-balloon. A relatively safe one, since it is sectional and on a matter not likely to arouse other sections of the nation to the support of the South. If it goes off fairly well, a precedent has been established. Govt by fiat can replace the Constitution. You don't have to credit me with too much intelligence and penetration, just so you watch carefully and think.

Meanwhile, personally, I am not delighted. I am not persuaded and elevated by the white mare technique. Negro schools in the state are in very good shape and on the improve. We are fortunate in having Dr. D. E. Williams as head and driving force of Negro instruction. Dr. Williams is relentless in his drive to improve both physical equipment and teacher quality. He has accomplished wonders in the 20 years past and it is to be expected that he will double that in the future.

It is well known that I have no sympathy nor respect for the "tragedy of color" school of thought among us, whose fountain-head is the pressure group concerned in this court ruling. I can see no tragedy in being too dark to be invited to a white school social affair. The Supreme Court would have pleased me more if they had concerned themselves about enforcing the compulsory education provisions for Negroes in the South as is done for white children. The next 10 years would be better spent in appointing truant officers and looking after conditions in the homes from which the children come. Use to the limit what we already have.

Thems my sentiments and I am sticking by them. Growth from within. Ethical and cultural desegregation. It is a contradiction in terms to scream race pride and equality while at the same time spurning Negro teachers and self-association. That old white mare business can go racking on down the road for all I care.

—ZORA NEALE HURSTON
Eau Gallie

POLITICAL CARTOONS

A highly expressive and at times provocative genre, political cartoons are at once visual and textual, artistic and political (at times exaggerated and realistic, humorous and sober). As a hybrid genre, they move in imaginative ways across the boundaries of varied vocabularies, such as art and politics, to make an argument. It is helpful to think of political cartoonists as cultural workers who artistically represent and perceptively interpret an issue. In other words, they vividly depict, or symbolize, a concern. The visual representation is typically illustrative as opposed to being abstract; the written interpretation is usually satiric — or a critical examination of human folly, vice, and evil — rather than approvingly descriptive. And the mode is often ironic, meaning that a serious issue is presented humorously, or vice versa. In the best political cartoons, the interplay between the textual and the visual accentuates the message.

The challenge here is to analyze how and to what effect the following political cartoons — which come from southern and northern, black and white newspapers — jointly represent and interpret *Brown*. Do you think, as many have argued, that the merger of art and politics in forms like political cartoons dilute their power as political as well as artistic statements? How are these political cartoons alike and how are they different? Do you detect any regional or racial patterns? Which of the political cartoons are more or less effective? Why? Finally, how viable are political cartoons as historical source material? Do they in any way tell us something about *Brown* as well as or even better than other kinds of materials?

CHRISTIAN SCIENCE MONITOR (BOSTON)

May 22, 1954

"No job for a race horse."

Little in The Nashville Tennessean

"The thinker."

A Supreme Court Bomb!

WHITE BACKLASH

The most important immediate consequence of *Brown* was the impetus it provided to the ongoing black freedom struggle. It helped spark the modern Civil Rights and Black Power movements. Likewise, it forced white Americans to confront more directly the depth and ubiquity of their investment in white supremacy. Some gave their active support to the intensifying black liberation struggle. Many adopted a wait-and-see attitude: They tacitly supported racial equality, especially when it did not interfere with their sense of economic entitlement or, ironically, their sense of white privilege.

An increasingly vocal contingent, notably in the white South, used *Brown* as a focus for their opposition to the black freedom struggle and egalitarianism on the one hand and their unswerving commitment to white supremacy on the other. Adopting the cloak of states' rights and privacy rights, these diehard segregationists mounted an increasingly vigorous opposition to *Brown* and what it signified: not just the end of Jim Crow, but the emergence of integrationism as the dominant cultural ideal.

What follows are two documents which capture aspects of southern white resistance to *Brown*. Highly influential die-hard segregationist senators Strom Thurmond of South Carolina and Harry Byrd of Virginia led the drafting and dissemination of "The Southern Manifesto." The southern white ruling elite's powerful opposition to *Brown* and racial integration helped fuel the white racist counterinsurgency against the evolving southern black civil rights insurgency. One consequence of this intense resistance, particularly in the face of the mounting black civil rights insurgency, was that southern school desegregation was effectively delayed until the courts intervened in the late 1960s. The other document is a cultural artifact: a handbill disseminated by the New Orleans chapter of the White Citizen's Council—a right-wing white supremacist organization—imploring parents to prevent their children from falling under the sway of 'Negro' music. What is the basic claim of the manifesto? On what grounds is that claim advanced? What do you see as the strengths and weaknesses of these positions? How do you account for the depth of southern white opposition to *Brown* and what it signified?

Do you see any demonstrable links between the handbill and the manifesto? Why do you think that music, especially black music, excited such concern among these parents? How might historians use sources like this handbill: what can it tell them? What are its strengths and weaknesses as a historical source?

The Southern Manifesto
March 12, 1956

Declaration of Constitutional Principles

The unwarranted decision of the Supreme Court in the public school cases is now bearing the fruit always produced when men substitute naked power for established law.

The Founding Fathers gave us a Constitution of checks and balances because they realized the inescapable lesson of history that no man or group of men can be safely entrusted with unlimited power. They framed this Constitution with its provisions for change by amendment in order to secure the fundamentals of government against the dangers of temporary popular passion or the personal predilections of public officeholders.

We regard the decision of the Supreme Court in the school cases as a clear abuse of judicial power. It climaxes a trend in the Federal judiciary undertaking to legislate in derogation of the authority of Congress, and to encroach upon the reserved rights of the States and the people.

The original Constitution does not mention education. Neither does the 14th amendment nor any other amendment. The debates preceding the submission of the 14th amendment clearly show that there was no intent that it should affect the systems of education maintained by the States.

The very Congress which proposed the amendment subsequently provided for segregated schools in the District of Columbia.

When the amendment was adopted, in 1868, there were 37 States of the Union. Every one of the 26 States that had any substantial racial differences among its people either approved the operation of segregated schools already in existence or subsequently established such schools by action of the same lawmaking body which considered the 14th amendment.

As admitted by the Supreme Court in the public school case *(Brown v. Board of Education)*, the doctrine of separate but equal schools "apparently originated in *Roberts* v. *City of Boston* . . . (1849), upholding school segregation against attack as being violative of a State constitutional guarantee of equality." This constitutional doctrine began in the North, not in the South, and it was followed not only in Massachusetts, but in Connecticut, New York, Illinois, Indiana, Michigan, Minnesota, New Jersey, Ohio, Pennsylvania, and other northern States until they, exercising

Congressional Record, 84th Cong., 2nd sess., March 12, 1956, 4459–64.

their rights as States through the constitutional processes of local self-government, changed their school systems.

In the case of *Plessy* v. *Ferguson* in 1896 the Supreme Court expressly declared that under the 14th amendment no person was denied any of his rights if the States provided separate but equal public facilities. This decision has been followed in many other cases. It is notable that the Supreme Court, speaking through Chief Justice Taft, a former President of the United States, unanimously declared in 1927 in *Lum* v. *Rice* that the "separate but equal" principle is "within the discretion of the State in regulating its public schools and does not conflict with the 14th amendment."

This interpretation, restated time and again, became a part of the life of the people of many of the States and confirmed their habits, customs, traditions, and way of life. It is founded on elemental humanity and commonsense, for parents should not be deprived by Government of the right to direct the lives and education of their own children.

Though there has been no constitutional amendment or act of Congress changing this established legal principle almost a century old, the Supreme Court of the United States, with no legal basis for such action, undertook to exercise their naked judicial power and substituted their personal political and social ideas for the established law of the land.

This unwarranted exercise of power by the Court, contrary to the Constitution, is creating chaos and confusion in the States principally affected. It is destroying the amicable relations between the white and Negro races that have been created through 90 years of patient effort by the good people of both races. It has planted hatred and suspicion where there has been heretofore friendship and understanding.

Without regard to the consent of the governed, outside agitators are threatening immediate and revolutionary changes in our public-school systems. If done, this is certain to destroy the system of public education in some of the States.

With the gravest concern for the explosive and dangerous condition created by this decision and inflamed by outside meddlers:

We reaffirm our reliance on the Constitution as the fundamental law of the land.

We decry the Supreme Court's encroachments on rights reserved to the States and to the people, contrary to established law and to the Constitution.

We commend the motives of those States which have declared the intention to resist forced integration by any lawful means.

We appeal to the States and people who are not directly affected by these decisions to consider the constitutional principles involved against

White Citizen's Council Notice.

the time when they, too, on issues vital to them, may be the victims of judicial encroachment.

Even though we constitute a minority in the present Congress, we have full faith that a majority of the American people believe in the dual system of Government which has enabled us to achieve our greatness and will in time demand that the reserved rights of the States and of the people be made secure against judicial usurpation.

We pledge ourselves to use all lawful means to bring about a reversal of this decision which is contrary to the Constitution and to prevent the use of force in its implementation.

In this trying period, as we all seek to right this wrong, we appeal to our people not to be provoked by the agitators and troublemakers invading our States and to scrupulously refrain from disorders and lawless acts.

[The names of the signers were affixed to the text — 19 from the Senate and 77 from the House of Representatives, making a total of 96 signatures.—ED.]

NATIONAL PROGRESS REPORT: REALIZING INTEGRATED SCHOOLS

Since the 1954 *Brown* decision, the struggles to achieve school desegregation and an integrated society have waxed and waned. Clearly, however, the promise of that momentous decision has not been realized. This nagging and deeply troubling national frustration has called forth a most revealing commemoration: anniversaries of the decision wherein the nation takes stock of its halting progress toward racial integration. What follows are the *New York Times*'s tenth-, twentieth-, thirtieth-, and fortieth-year editorial report cards. These sorts of ritualistic national bows before the altar of *Brown* have yielded much impressive rhetoric throughout the country. Unfortunately, they have not been able to tap into or help create the kind of national resolve required to meet the serious challenges posed by a commitment to an integrated society and two of its basic premises: integrated neighborhoods and schools.

Do you see any recurring themes in these editorials? Are there salient similarities and differences among them? How do they compare with the *Times*'s 1954 editorial on *Brown* (p. 208)? Where are we today in our struggle to realize the promise of *Brown?*

Decade of Desegregation

May 17, 1964

On its tenth anniversary the unanimous decision of the Supreme Court outlawing racial segregation in the public schools stands as the great turning point in the battle for civil rights. The battle is still far from won—in the schools, in voting, in jobs, in housing or in any other major aspect of American life—but the commitment to equal opportunity is irrevocable, the outcome certain.

The questions lie in the pace and pains of progress. How swiftly will we complete the arsenal of laws required to provide effective safeguards against racial discrimination? And, more fundamental still, how fully, how peacefully and how fast will we accomplish the transformation of attitudes necessary to make equality real in every community, North and South?

On the legislative front, Senate leaders of both parties will be trying this week to muster enough votes to break the Southern filibuster and to assure passage of the civil rights bill.[1] Confidence in the Congressional process will be maintained best if the Senate acts promptly without sacrificing any basic provisions of the House-passed measure as the price of final approval.

After that will come the infinitely more complex task of white and Negro adjustment to translating legal equality into the day-to-day practices of community life. The history of school desegregation makes plain how difficult this adjustment is likely to be. In the decade since the Supreme Court struck down legally enforced racial separation, official desegregation has taken place in only 423 of the 2,256 Southern school districts with Negro and white children. And even this exaggerates the extent of integration. A survey by the Southern Education Reporting Service indicates that only 34,110 of the South's 2,900,000 Negro pupils actually attend school with white children.

In the North, where Negro ghettos produce a racial separation almost as rigid as any in the South, community conflict over methods of speeding integration has taken on an increasingly ugly aspect in many large cities. Education is the key to full emancipation for the Negro; its impor-

---

[1]The 1964 Civil Rights Act was a watershed: the most powerful series of federal civil rights initiatives ever undertaken in areas such as education, public accommodations, and voting. The 1965 Voting Rights Act solidified the federal initiative in that arena.

tance is vastly enhanced in this period when automation is taking its heaviest toll in the unskilled and semi-skilled jobs that have up to now been the Negro's primary economic reliance.

Making available enormously improved facilities for overcoming the educational handicaps of children reared in the slums is an urgent part of the job of building truly open communities, in which there will be no second-class citizens. The war on poverty and the still lagging campaign for full employment are also key ingredients. So, too, is the practice of working and studying together and living in harmony as neighbors. The more experience we have in making equality work, the less need there will be for litigation and the less opportunity for extremists to foster race hatred and violence.

The original decision of the Supreme Court has significantly broadened the role of the judiciary in the defense of human rights; it has caused a substantial shift in the balance of Federal-state relations; it has involved the Court itself in controversy of an intensity it has not known since the early years of the New Deal. Once under attack as too conservative, the Court now finds its critics chiefly on the right—among those who consider it too radical an architect of social change. In the perspective of history, its school decision will rank among the most momentous the Court ever made. But its real implementation lies with the people.

TIMES (NEW YORK)

Two Decades Later

May 17, 1974

Twenty years ago, Chief Justice Earl Warren wrote for a unanimous Supreme Court in *Brown v. Board of Education:* "We conclude that in the field of public education the doctrine of 'separate but equal' has no place." The nine Justices agreed that to separate black children by order of the law "may affect their hearts and minds in a way unlikely ever to be undone."

May 17, 1954, marked more than the start of a laborious dismantling of the South's dual public education systems. The ruling signaled the end of Jim Crow, the segregationist doctrine sanctioned by the disastrous Supreme Court decision in *Plessy v. Ferguson* in 1896. Ignoring Justice

John Marshall Harlan's prophetic dissenting view that "our Constitution is color blind, and neither knows nor tolerates classes among citizens," the Court had codified post-Reconstruction racism by giving the "separate but equal" doctrine the status of substitute for "equal protection."

Unrelated to education, *Plessy* had upheld segregated seating on public streetcars. Yet, for almost sixty years that judicial abomination provided the constitutional basis for the South's apartheid. It was a doctrine that infected the nation far beyond Jim Crow's official borders.

In catching up with Justice Harlan's dissent, the Warren Court restored the Constitution's integrity. Despite much resistance, *Brown* set in motion an irreversible social revolution. The debate over the decision's impact too often is confined to statistics of school desegregation.

Within a year of *Brown,* Rosa Parks, a tired seamstress in Montgomery, Alabama, was, like Homer Plessy sixty years earlier, arrested for her refusal to move to the back of a bus. A little known minister named Martin Luther King Jr. brought the public company to its knees by keeping blacks off its buses for more than a year.

In Little Rock, President Eisenhower ordered troops to escort children to school past a human wall of segregationists. Black students in North Carolina occupied segregated lunch counters in nonviolent protest until the illegal barriers fell.

Blacks and whites marched and fought together, daring the guns, dogs and obscenities of white sheriffs. An army of aroused Americans marched to Washington to pledge support for Dr. King's dream of equality.

Great universities dropped their restrictive color bars.

Finally, in 1964 and 1965, prodded by President Johnson—to his eternal credit—Congress enacted the civil rights and voting rights laws that dramatically changed the roles of blacks in employment, the electoral process and the political power structure.

The twenty-year march was slowed by serious setbacks—the murder of Dr. King, the explosions of the urban ghettos, the tactics of white segregationists and black separatists. President Nixon has persisted in trying to negate the Constitution with divisive anti-busing appeals and proposals, the latest of which was narrowly defeated by the Senate this very week.

The wounds of racial hatred have not yet healed. Old suspicions and new economic fears still divide races and classes. And yet, the nation is moving irrevocably toward its integrated goal under a Constitution that is in fact color blind. The dual school systems are no more. Black mayors have been elected in great cities, including the South. The public schools in two of the most populous states — California and Michigan — are headed by black educators. Among the Justices of the Supreme Court is Thurgood Marshall, the lawyer who argued the case for the reversal of *Plessy* before the Warren Court.

Today's anniversary of segregation's historic defeat calls not for self-congratulatory paeans but for a pledge to build on the foundation of considerable but insufficient gains, with renewed faith and with more than deliberate speed.

TIMES (NEW YORK)

The Enduring Promise of Brown

May 17, 1984

Anniversaries of Supreme Court decisions don't usually inspire celebration. But nothing less is in order this week, the 30th anniversary of the decision by which the Court struck down its own colossally wrong acceptance of "separate but equal" treatment for blacks and whites in the preceding half century. To celebrate *Brown v. Board of Education* is to celebrate a continuing revolution in America's race relations.

The *Brown* decision was evoked by a carefully developed, 20-year legal assault on racial segregation in public education. But its logic, that separate is inherently unequal, reached far beyond the classroom and still reverberates throughout American society.

Until Brown, the painful three-century struggle of blacks to move from slavery to full citizenship seemed to be obstructed by the Constitution itself. Despite great cultural advance, generations of blacks could see no end of segregation. But on May 17, 1954, all perspectives changed.

The Brown ruling mandated desegregation in public education — democracy's most promising ticket to equality. It also signaled the end of Jim Crow and the badge of inferiority that had been "legally" imposed on all black Americans.

And the decision has stood rock solid as a matter of law, despite the continuing resistance, not only in the South, to desegregation in schools and housing, and despite the lukewarm enforcement efforts of several

Administrations. *Brown* has left no doubt about the right even where conditions are still woefully wrong.

Even the worst Jim Crow laws did not collapse without further struggle. Segregation in buses and restaurants and voting booths had to be challenged and resisted with boycotts, sit-ins, marches and other demonstrations, almost all nonviolent.

But since *Brown,* the blacks and whites who fought those battles have had a proud and legal banner to display. Within a decade and a half, *Brown*'s principles were finally written into laws that prohibit discrimination in public accommodations, employment and housing.

The most important of those laws by far was the 1965 Voting Rights Act, whose fruits are only now ripening. With the vote secured, blacks have won a growing number of political offices and gained political awareness and strength, as can be seen from their rallying to the Presidential campaign of the Rev. Jesse Jackson.

Brown v. Board of Education stands as a national confession of error, a true landmark. It propelled the modern civil rights movement, a still-incomplete social revolution. It reaffirmed the American spirit of equality and rekindled hope of peaceful transformation. It is a living monument, a cause for celebration.

TIMES (NEW YORK)

Forty Years and Still Struggling

May 18, 1994

Yesterday's celebration of the 40th anniversary of *Brown v. Board of Education,* the landmark school desegregation decision, was an occasion both for national pride and national shame. In 1954 the Supreme Court knocked over the bankrupt notion of separate but equal. Chief Justice Earl Warren, guiding the Court to unanimity with both moral fervor and political sense, liberated the nation from centuries of apartheid.

Forty years later, it seems the national consciousness has changed but the reality lags far behind. An unacceptable number of minority students still attend all-minority or nearly all-minority schools. Where statutorily sanctioned segregation once kept the races apart at school, now the trends of suburbanization, white flight, industrial decay and political ennui have combined to keep minority students in poorer districts, in schools with lower expectations for their students' achievement, in communities that are nurseries for failure.

The Supreme Court said in 1896 that separate seating on public transportation did not deny equality to anyone. It fatuously argued that if enforced segregation stamped one race as inferior, that was "solely because the colored race chooses to put that construction upon it."

America could not begin to move forward before rejecting that bogus sociology. The Chief Justice held simply that separate facilities were "inherently unequal" in education, and later in transportation and other aspects of life.

That, however, only began a new chapter of the struggle that met widespread resistance in the South and more subtle non-compliance in the North. The goal of genuinely integrated schooling suffered a tragic setback in 1974 when the Supreme Court, in a case from Michigan, allowed the city line across which many whites had fled to set the boundaries of Detroit's efforts to desegregate its schools.

Still, the school decision slowly awakened America's conscience and energized Congress to topple segregation in other areas, most conspicuously in public accommodations, where anticipated resistance quickly melted away. While legal employment barriers also were hurdled, many minorities still see only a widening income gap in a shrinking economy.

Another conspicuously successful product of *Brown*, the 1965 Voting Rights Act, brought thousands of minorities to elective office. Indeed, President Clinton told the pupils of Martin Luther King Jr. Middle School in Beltsville, Md., yesterday that his own election was due in large part to the support of empowered minorities.

The anniversary's blend of satisfaction and frustration would not have surprised Thurgood Marshall, strategist of the movement that won the *Brown* case. He and his colleagues in the NAACP Legal Defense Fund were under no illusions of immediate success.

The struggle continues. Lawsuits all around the country are now challenging the inherent inequities in the funding of school districts with local property taxes—creating huge disparities in the resources available to children in city and suburban districts—and in the district lines themselves, which allow residential segregation to be reflected in the schools.

In the decade between this year and the 50th anniversary of this revolutionary decision, Americans owe it to their children to find ways to live up to its promise. If they do not, the price will be high.

Epilogue:
The Legacy of *Brown*

In a penetrating assessment of black participation in Reconstruction politics, W. E. B. Du Bois judged those significant individual and collective efforts to substantiate black freedom "a splendid failure." He explained that whites thought Reconstruction would fail because of innate black incompetence. The exact opposite happened, however. Black politicians exercised their rights and shouldered their responsibilities competently; in fact, they acquitted themselves as well as, and often better than, their white cohorts. Rather than confirming black incapacity, black Reconstruction confirmed human equality. This exemplary illustration of black achievement in spite of powerful white opposition, Du Bois maintained, was "splendid."

Nevertheless, Reconstruction failed for a variety of reasons. Most important, the federal government did not provide the ex-slaves with land—an economic stake in society and the necessary material basis to complement their newfound political and civil rights. It also failed because of the white counterinsurgency and the resulting institutionalization of white supremacy that we know as disfranchisement, antiblack terrorism and violence, and Jim Crow. Racist white opposition to and backlash against black progress—real and imagined—has been a common recurrence in the history of American race relations. Economic downturns and interracial economic competition on one hand and the perception that black progress has come at the expense or behest of whites on the other have been central to this ongoing pattern.

Many observers have characterized the modern black freedom struggle (1955–75) as the second Reconstruction. A common assessment of this enormously influential mass movement stresses—like the prevailing wisdom regarding the first Reconstruction—a mixed or bittersweet legacy. Some adopt the Du Boisian notion of a "splendid failure" as an equally persuasive appraisal of the second Reconstruction. Legal and

230

constitutional change—as exemplified by the Thirteenth, Fourteenth, and Fifteenth Amendments and the 1875 Civil Rights Law—was pivotal to the first Reconstruction. Likewise, such change, as embodied in the *Brown* decision, has proven vital to the second.

Any assessment of the historical legacy of *Brown* must begin with the acknowledgment that thus far it has been complex, contested, and at times ambiguous. To date, the meanings and consequences of that legacy have been many, conflicting, and highly dependent on the observer's angle of vision. Many conservatives have viewed *Brown* and its implementation as federal trampling on both states' rights and white racial privilege. Blacks and their progressive allies, on the contrary, have tended to view it as a necessary step toward squaring America's treatment of blacks with the American creed: a simple matter of justice and morality.

While *Brown* ended legalized school segregation, subsequently there has not developed a national consensus as to exactly what integrated schools and their even more important corollary, an integrated society, are and how they might be achieved. As a result, the realization of *Brown,* "with all deliberate speed," has been checkered at best. Its history has reflected the difficulties and hard choices involved in struggling to implement its mandate. Thus while *Brown* ended legalized Jim Crow in public school education, it did not end untold varieties of voluntary and actual racial segregation. Similarly, *Brown's* haphazard and varying nationwide implementation has not yielded the racially integrated elementary and secondary schools or equality of educational opportunity envisioned in the flush of its immediate afterglow. Likewise, in spite of federal mandates that the dual (separate white and black) college and university systems in various southern states be integrated, racially identifiable institutions persist: better-funded predominantly white ones and less well-funded historically black colleges and universities.

Paradoxically, the most integrated elementary and secondary schools nationally are in the very South where court-ordered desegregation proved necessary. Here school integration has worked best. Ten years after *Brown,* voluntary compliance in the South was a failure: 98 percent of blacks attended rigidly segregated schools. In 1997, after thirty years of federal mandates and local initiatives, that figure is approximately one-third. The most racially segregated schools in 1997 are in the North and West where metropolitan desegregation—characteristically busing across nonwhite urban and white suburban borders to achieve racial balance—has met a brick wall of white resistance. In fact, the most intensely racially segregated schools are in Illinois, New York, and New

Jersey, with their inner-city communities of color surrounded by white suburbs—as in Chicago, New York City, and Newark.[1]

Indeed, many have noted with alarm the growth in the 1990s of the segregation and resegregation of public schools throughout the country. These deeply disturbing developments have been sustained by the courts. In 1995, a Connecticut judge determined that the state bore no responsibility for the glaring disparities between Hartford's poorly funded, low-achieving overwhelmingly Latino and African American schools and the lavishly funded, high-achieving surrounding white sub-urban schools. The correlations between poverty and diminished academic achievement as well as intensely segregated schools and limited access to networks pivotal to mainstream success are ignored.

Resegregation is occurring where courts have allowed districts—like Norfolk, Virginia, in 1987—to return to neighborhood schools, residential segregation notwithstanding. In spite of its history of legalized seg-regation, Norfolk successfully lobbied for this "relief" as a way to enhance educational opportunity and achievement, notably for its African Ameri-can students. Ten years later that promise had not been met. Much like the late-nineteenth-century courts' quick retreat from the civil rights and voting rights initiatives of the first Reconstruction, their late-twentieth-century counterparts have rapidly retreated from their brief pro–school integration posture of the late 1960s and early 1970s.[2]

Riding the Black Power crest of the black freedom wave of the late 1960s and pushed by widespread southern white opposition to effective school desegregation strategies, the Court fashioned a series of rulings that led to modest progress toward integrated schools, especially in the South. *Green v. County School of New Kent County* (1968) demanded that segregated or dual systems be fully replaced with integrated or unitary systems. A series of criteria was outlined to assess progress toward this goal. Faculty, staff, and extracurricular activities, for example, had to be totally integrated to comply with the mandate in *Green*. The Mississippi case *Alexander v. Holmes County Board of Education* (1969) reinforced the demand in *Green* that unitary school systems be developed with dis-patch.

Even more important for the realization of school desegregation plans was the decision in the North Carolina case of *Swann v. Charlotte-Mecklenberg Board of Education* (1971). Invalidating ineffective and appar-ently "race-neutral" student assignment plans like the popular and mis-

[1] Gary Orfield, Susan Eaton, et al., *Dismantling Desegregation: The Quiet Reversal of Brown* v. Board of Education (New York: New Press, 1996), 58–59, 64–65.
[2] Ibid. This text is a compelling treatment of these and related issues.

leading "freedom of choice," the Court saw such options as reinforcing segregated schools by building on residential segregation. *Swann* called for the aggressive implementation of district-wide desegregated schools. In a move with powerful consequences, the Court sanctioned busing as a remedy to achieve integrated schools. This remedy continues to be highly controversial.

For the first time, *Keyes v. Denver School District No. 1* (1973) expanded the desegregation mandate to a school system outside the South with no history of school segregation. Holding the district responsible for school board policies and decisions (such as building schools in isolated communities of color), the Court ruled that Latinos as well as blacks had been discriminated against and, as a result, steps to achieve a unitary system had to be taken forthwith. These initiatives marked the high tide of judicial activism on behalf of school desegregation.

The conservative tilt of the Court since the early 1970s reflects an increasing white backlash against Black Power activism, numerous black urban insurrections—or riots—of the mid- to late 1960s, and the misperception that federally sponsored initiatives like jobs programs aided blacks at the expense of whites. The ensuing white counterinsurgency— much like that of the first Reconstruction—has sought to arrest the wheels of racial progress, especially in the arena of school desegregation.

Nowhere is this better illustrated than in the "massive resistance" of whites outside the South to school desegregation—particularly to the busing of white schoolchildren to achieve integration—in communities like Boston, Chicago, and Los Angeles. The national spread and coalescence of white opposition to school busing reveal a larger pattern of white opposition to an integrated and equitable society. Vigorous, deep-seated, and ubiquitous white hostility to equal employment opportunity, residential integration, as well as school desegregation are interrelated. In fact, systemic patterns of antiblack economic, political, and social discrimination have marked American life in all regions, not just southern life. Viewed in this light, it is not surprising that the racist white Alabaman George Wallace garnered a significant measure of support among western and northern whites for his presidential bids in 1968 and 1972.

The centrist Democratic Party politics of the last quarter of the twentieth century have too often reinforced rather than challenged the antiprogressive and antiblack conservatism of the modern Republican Party's ascendancy. The age of Richard Nixon (1968–74) and, in some ways even more terrifyingly, the age of Ronald Reagan (1980 to at least 1996), have exemplified white resistance to black progress signified by increasing opposition to race-based affirmative action programs as well as school desegregation. Ultimately, however, the race relations record has been profoundly mixed. For as the black middle class has expanded greatly since the late 1960s, so has the black hardship index — growing numbers of black children in poverty; poor, single, black female heads of households with children; and, black men in jails (as opposed to schools).

Set within this history, the Court's retreat from school desegregation can be seen in large part as a national backing away from the spirit of *Brown.* As witnessed after the first Reconstruction, after the second there has been a national waning of idealism and hope regarding racial integration as well as commitment to it. This waning has been part of a larger pattern of decline and loss. This can be seen in the loss of American worldwide hegemony, signaled by the economic dropoff beginning with the oil crisis of the early 1970s and the emergence of competitive economies, especially in Japan and West Germany. Also, the American defeat in the Vietnam War was a blow to the national ego and international stature. Furthermore, it can be glimpsed in the national unwillingness thus far to direct to domestic priorities such as education any of the purported economic savings from America's apparent victory in the cold war.

Rather than focusing on issues of economic and social justice, as the spirit of *Brown* called for, the nation has turned its attention to business and economic concerns. Much as those interests dominated the late-nineteenth-century United States — the era of Jim Crow's emergence — they dominate the late twentieth century. This latter era has likewise witnessed a parallel racist counterinsurgency, as evidenced in the popularity of the thinly veiled racist diatribes of Charles Murray and Richard Herrnstein in *The Bell Curve* (1994) and Dinesh D'Souza in *The End of Racism* (1995). In many ways, the Court's dwindling commitment to school desegregation has mirrored this backlash. *Milliken v. Bradley* (1974) effectively stymied efforts to create unitary school systems across white suburban and nonwhite urban lines by establishing a virtually impossible burden of proof for the plaintiff. It must be shown that the state or suburb intentionally pursued segregation. In *Milliken v. Bradley II* (1977), the Court, searching for a remedy consistent with *Milliken I,* ruled that the state could be ordered to subsidize compensatory educational initiatives to alleviate segregation, such as reduced class sizes, computer labs, and math and reading specialists. The results of this surface tinkering, in lieu of structural changes, have not been encouraging.

In *Missouri v. Jenkins* (1995), the Court declared that the remedies under *Milliken II* should have a time limit and a narrowly yet fully articulated goal. An overriding concern in these matters was the return of jurisdiction to local authorities. In fact, the Court determined that it was not even necessary to present evidence that segregation had actually been alleviated. A series of related Court decisions — *Board of Education of Oklahoma v. Dowell* (1991) and *Freeman v. Pitts* (1992) — have given added comfort to districts desiring to dismantle their desegregation programs. The former ruling stated that once a district had achieved formal recognition as integrated — contradictory evidence of segregation notwithstanding — the district could be absolved of its desegregation mandate. The latter determined that a district's desegregation mandate could be eased even if progress toward desegregation was decidedly mixed.

A narrow focus on a negative, instrumental, and legal view of *Brown*'s legacy is inadequate and potentially misleading, however. The significant indices of racial progress owing to *Brown*'s impact — including the phenomenal growth in the numbers and influence of local, state, and national black elected officials — demonstrate an increasing and remarkable measure of black empowerment. Aggregate black wealth is greater than it has ever been; more blacks are doing better than ever before. Paradoxically, the coexistence of black poverty and black progress — and their fundamental interrelationship — is a most revealing index of the prospects and perils of capitalism. There are always winners and losers within a capitalist economy. Racism further complicates the situation by hounding the black haves as well as the black have-nots. For instance, where property taxes are the basis for school funding formulas, the schools of the poor, especially those serving inner-city minorities, receive far less money. In *San Antonio Independent School District v. Rodriguez* (1973), the Court approved these property-tax-based school funding formulas, eroding the notion that poor children of color have a right to equal educational opportunity, never mind economic justice.

The legacy of *Brown,* it must be understood, is cultural as well as legal. Consequently, it is imperative that we consider how that momentous ruling has expanded and enriched not just our civic culture but our national culture in its various and sundry manifestations, from the popular to the elite. *Brown* tapped deeply into the fundamental national sense of the United States as a community of shared beliefs and values. It embodies the best of our hopes and ideals. It resonates with our best selves and our highest and most honorable callings. It springs from our bedrock national commitment to freedom, justice, and equality. In a sense, *Brown* is a metaphor for the American dream.

Again, this does not mean that the struggle is at all easy; often the exact opposite is the case. In a society where race matters profoundly, competing claims regarding what constitutes notions like justice and equality make agreeing upon and realizing them exceedingly difficult. Not surprisingly, therefore, the Court ruling in *United States v. Fordice* (1992) confirmed that Mississippi continued to operate a dual system of colleges in defiance of the federal mandate to desegregate. Here *Brown* was clearly being extended to colleges and universities. Unfortunately, the Court provided no guidance about how to deal with the effects of the continuing race-based inequities plaguing the system. Difficult questions persist, notably in light of similar situations in other southern states. If, as the Court argued, parity is not the answer, what is? In addition, can historically black colleges and universities—given their vital role in black education, their significance as black institutions, and their central role as black cultural spaces—exist within a truly integrated system? Does integration necessarily mean the end of both distinctive black institutions and distinctive black cultures? Otherwise stated, is integration consistent with multiculturalism?

Yes, for *Brown* signifies an open, tolerant, diverse, democratic, and inclusive national culture. It bears reiteration that this vision is being constantly contested and that it is simultaneously a process and a goal. Still, the extraordinary power and appeal of this vision continues, and it gives hope to those who value an integrated society. Rooted in cultural pluralism—respect for differences across group-based identities (race, ethnicity, gender, class)—it seeks national unity and national identity with its emphasis on integral ties to a community of shared values and beliefs.

To envision a future legacy for *Brown* as a "splendid failure" is to sell ourselves woefully short. We can and must do better. It is clear that the promise of equal educational opportunity in *Brown* demands proactive measures as well as renewed commitment to an integrated society. Our brief recent national experience with school desegregation confirms its superiority to our traditional commitment to segregation and exclusion. School desegregation cannot possibly succeed, however, unless it is linked with firm national resolve and political will. This means that residential integration must replace segregated housing patterns, and, even more important, job opportunities must be equalized. In light of the history and continuing evidence of racial discrimination, affirmative action is imperative if these patterns are to be alleviated. Ultimately, a color-blind emphasis on formal equality or mere equality of opportunity is grossly insufficient. The enduring challenge of *Brown* demands that we recommit ourselves as a nation to substantive, material equality: to an equality of results. Our vast resources, in particular our wealth and our structures of opportunity, must be more equitably distributed. Otherwise the legacy of *Brown* will indeed become frozen as another "splendid failure." We will again have failed the challenge of realizing a splendid success. This time we will have reneged on *Brown*'s future promise.

Chronology of Events Related to
Brown v. Board of Education

1793 U.S. Congress adopts the first Fugitive Slave Law, which increases the possibilities for the extradition of slaves and makes it a criminal offense to protect a fugitive slave.

1822 Denmark Vesey, a free black, organizes slave rebellion in Charleston, S.C.

1831 Nat Turner rebellion.

1849 *Roberts v. City of Boston* declares separate black and white schools legal (later overturned by state law).

1850 Compromise of 1850 strengthens 1793 Fugitive Slave Law.

1857 *Dred Scott* decision denies U.S. citizenship to African Americans.

1861 Civil War begins.

1863 Emancipation Proclamation.

1865 Civil War ends and Reconstruction begins; the Thirteenth Amendment ends slavery.

1868 The Fourteenth Amendment defines citizenship to include blacks.

1870 The Fifteenth Amendment gives blacks the right to vote.

1873 *Slaughterhouse Cases* narrowly define federal power and emasculate the Fourteenth Amendment by asserting that most of the rights of citizens remain under state control.

1877 Reconstruction ends.

1883 *Civil Rights Cases* invalidate provisions of the Civil Rights Act of 1875 and declare that the Fourteenth Amendment does not prohibit discrimination by private individuals or businesses.

1890 Louisiana passes a Jim Crow law mandating "separate but equal" accommodations on railroads for whites and blacks.

1892 Homer Plessy is arrested in Louisiana for riding in a white car on an intrastate trip.

1896 In *Plessy v. Ferguson,* the Supreme Court upholds the constitutionality of Louisiana's separate but equal law.

238

1899   In *Cumming v. Richmond County Board of Education,* the Supreme Court rules that public school education is within the purview of the states rather than the federal government.

1908   In *Berea College v. Kentucky,* the Supreme Court rules that private educational institutions must abide by the segregation laws of the state.

1909   NAACP is founded.

1917   *Buchanan v. Warley,* early NAACP-led legal victory, outlaws a municipal ordinance mandating residential segregation.

1927   *Gong Lum v. Rice* upholds the right of the state to define and enforce racial classifications for educational purposes.

1933   NAACP begins to attack segregation and discrimination in education at the graduate and professional levels through legal suits; its first case is a suit against the University of North Carolina on behalf of Thomas Hocutt, which was lost on a technicality.

1934   Charles Hamilton Houston appointed NAACP special legal counsel.

1936   In *University of Maryland v. Murray,* the Maryland Supreme Court orders that a black student be admitted to the state's white law school; Thurgood Marshall joins NAACP legal staff.

1938   In *Missouri ex rel. Gaines v. Canada,* the Supreme Court orders that Lloyd Gaines, an African American resident, be admitted to Missouri's all-white law school.

1939   Under the leadership of Charles Houston, the NAACP Legal Defense and Educational Fund established.

1944   Gunnar Myrdal's *An American Dilemma* published.

1950   In *Sweatt v. Painter* and *McLaurin v. Oklahoma State Regents,* the Supreme Court makes clear that the separate but equal standard in state-supported higher education is unattainable; Charles Houston dies.

1952   Initial oral arguments in *Brown* before Supreme Court.

1953   Reargument in *Brown* regarding original intent and possible relief.

1954   In *Brown v. Board of Education (Brown I)*, separate black and white schools declared illegal; *Plessy* overturned.

1955   *Brown II* mandates that schools be desegregated "with all deliberate speed."

1955   Emmett Till lynched.

1955–56   Montgomery bus boycott to alleviate antiblack discrimination on city's bus lines succeeds; local preacher and leader Martin Luther King Jr. emerges as the major leader of the African American freedom struggle.

1957   President Eisenhower sends the National Guard to Little Rock, Arkansas, to escort nine black students to Central High School.

1960   Four black students from Greensboro, N.C.'s A & T College sit-in to integrate Woolworth's lunch counter initiating a wave of similar — ultimately successful — protests throughout the South.

1961   Freedom Rides, led by the Congress of Racial Equality, to test compliance in the Deep South with the Court ruling outlawing discrimination in interstate bus terminals leads to several ugly incidents.

1963   Birmingham campaign of marches, sit-ins, and boycotts to get rid of segregation and discrimination throughout city incurs vicious local police response, including the hosing of schoolchildren and the jailing of King. National fallout contributes to passage of 1964 Civil Rights Act.

1964   Civil Rights Act of 1964 bans discrimination in voting and public accommodations; fair employment practices mandated.

1965   Violent repression of Selma; march to highlight continuing exclusion of black voters helps push through the 1965 Voting Rights Act. It nullifies literacy and other voting restrictions; federal authorities are empowered to register blacks to vote in various southern counties; a skirmish between local blacks and the Los Angeles Police Department escalates into Watts rebellion, an extraordinary racial disturbance which exacts a terrible toll in deaths and injuries, and roughly $225 million in property damage.

1967   Black rebellions in Newark, New Jersey, and Detroit, Michigan; Thurgood Marshall becomes first black named to Supreme Court.

1968   King murdered; Fair Housing Act passed, banning discrimination in the sale and leasing of houses and apartments; *Green v. County School Board of New Kent County* (Virginia) mandates that segregated school systems must be desegregated "root and branch."

1971   *Swann v. Charlotte-Mecklenburg Board of Education* disallows desegregation plans that are "race neutral" in design but reinforce segregation because of residential segregation; busing approved as a desegregation remedy.

1974   *Milliken v. Bradley* forbids city-suburban desegregation plans unless state or suburb is shown to be liable for urban segregation; extreme burden of proof for plaintiff effectively curtails interdistrict, urban-suburban integration plans.

1977   *Milliken v. Bradley II* empowers courts to demand that state pay for compensatory educational programs to alleviate harm caused by segregation.

1986   *Riddick v. School Board of the City of Norfolk, Virginia,* first federal ruling allowing a school district declared desegregated to revert to local control and neighborhood schools by getting rid of its desegregation plan.

1992   *United States v. Fordice* upholds mandate that Mississippi must dismantle its dual system of colleges and universities without guidance as to how this should be done.

Selected Bibliography

Anderson, James D. *The Education of Blacks in the South, 1860–1935.* Chapel Hill: University of North Carolina Press, 1988.

Auerbach, Jerold. *Unequal Justice: Lawyers and Social Change in Modern America.* New York: Oxford University Press, 1976.

Bardolph, Richard. *The Civil Rights Record: Black Americans and the Law, 1849–1970.* New York: Crowell, 1970.

Bass, Jack. *Unlikely Heroes: The Dramatic Story of the Southern Judges of the Fifth Circuit Who Translated the Supreme Court's* Brown *Decision into a Revolution for Equality.* New York: Simon and Schuster. 1981.

Belknap, Michael R., ed. *Desegregation of Public Education.* Vol. 7 of *Civil Rights, the White House, and the Justice Department.* New York: Garland, 1991.

Bell, Derrick. *Race, Racism, and the Law.* 3rd ed. Boston: Little, Brown, 1992.

Bell, Derrick, ed. *Shades of Brown: New Perspectives on School Desegregation.* New York: Teachers College, Columbia University, 1980.

Berry, Mary Frances. *Black Resistance — White Law: A History of Constitutional Racism in America.* New York: Penguin Press, 1994.

Blaustein, Albert P., and Clarence Clyde Ferguson Jr. *Desegregation and the Law: The Meaning and Effect of the School Segregation Cases.* New Brunswick: Rutgers University Press, 1957.

Boxill, Bernard R. *Blacks and Social Justice.* Totowa, N.J.: Rowman and Allanheld, 1984.

Brown, Oliver, et al., Appellants. *Brief for Appellants in Nos. 1, 2, and 4 and for Respondents in No. 10 on Reargument.* In the Supreme Court of the United States, October Term, 1953.

Browning, R. Stephen, ed. *From Brown to Bradley: School Desegregation, 1954–1974.* Cincinnati: Jefferson Law Book Company, 1975.

Chesler, Mark A., Joseph Sanders, and Debra S. Kalmuss. *Social Science in Court: Mobilizing Experts in the School Desegregation Cases.* Madison: University of Wisconsin Press, 1988.

Crain, Robert L. *The Politics of School Desegregation: Comparative Case Studies of Community Structure and Policy-Making.* Chicago: Aldine, 1968.

Fancher, Betsy. *Voices from the South: Black Students Talk about Their Experiences in Desegregated Schools.* Atlanta: Southern Regional Council, 1970.

Finkelman, Paul, ed. *The Struggle for Equal Education*. Pt. 1, Vol. 7 of *The African-American Experience*. New York: Garland, 1992.

Flicker, Barbara, ed. *Justice and the School Systems: The Role of Courts in Education Litigation*. Philadelphia: Temple University Press, 1990.

Franklin, John Hope, and Genna Rae McNeil, eds. *African Americans and the Living Constitution*. Washington, D.C.: Smithsonian Institution Press, 1995.

Friedman, Leon, ed. *Argument: The Oral Argument before the Supreme Court in Brown v. Board of Education of Topeka, 1952–55*. New York: Chelsea House, 1969.

Graglia, Lino A. *Disaster by Decree: The Supreme Court Decisions on Race and the Schools*. Ithaca: Cornell University Press. 1976.

Greenberg, Jack. *Crusaders in the Courts*. New York: Basic, 1994.

Guzman, Jessie P. *Twenty Years of Court Decisions Affecting Higher Education in the South, 1938–1958*. Tuskegee, Ala.: Tuskegee Institute Press, 1960.

Harlan, Louis R. *Separate and Unequal: Public School Campaigns and Racism in the Southern Seaboard States, 1901–1915*. 1958; reprint, New York: Atheneum, 1968.

Higginbotham, A. Leon. *In the Matter of Color: Race and the American Legal Process, the Colonial Period*. New York: Oxford University Press, 1978.

Hochschild, Jennifer L. *The New American Dilemma: Liberal Democracy and School Desegregation*. New Haven: Yale University Press, 1984.

———. *Thirty Years after Brown*. Washington, D.C.: Joint Center for Political Studies, 1985.

Horowitz, Harold K., and Kenneth L. Karst. *Law, Lawyers and Social Change: Cases and Materials on the Abolition of Slavery, Racial Segregation and Inequality of Educational Opportunity*. Indianapolis: Bobbs-Merrill, 1969.

Jones, Leon, ed. *From Brown to Boston: Desegregation in Education, 1954–1974*. Vol. 1, *Articles and Books*. Metuchen, N.J.: Scarecrow Press, 1979.

———. *From Brown to Boston: Desegregation in Education, 1954–1974*. Vol. 2, *Legal Cases and Indexes*. Metuchen, N.J.: Scarecrow Press, 1979.

Karst, Kenneth L. *Belonging to America: Equal Citizenship and the Constitution*. New Haven: Yale University Press, 1989.

Kirp, David L. *Just Schools: The Idea of Racial Equality in American Education*. Berkeley: University of California Press, 1982.

Klarman, Michael J., et al., *"Brown, Racial Change, and the Civil Rights Movement," Virginia Law Review* 80, no. 1 (Feb. 1994), 7–150.

Kluger, Richard. *Simple Justice: The History of Brown v. Board of Education and Black America's Struggle for Equality*. New York: Knopf, 1975.

Kousser, J. Morgan. *Dead End: The Development of Nineteenth-Century Litigation on Racial Discrimination in the Schools*. Oxford: Clarendon Press, 1986.

Kurland, Philip B., and Gerhard Casper, eds. *Landmark Briefs and Arguments of the Supreme Court of the United States: Constitutional Law [Brown v. Board of Education (1954–1955)]*, vols. 49, 49A. Arlington, Va.: University Publications of America.

Levin, Betsy, and Willis D. Hawley, eds. *The Courts, Social Science, and School Desegregation*. New Brunswick, N.J.: Transaction Books, 1977.

Levy, Leonard W., and Douglas L. Jones, eds. *Jim Crow in Boston: The Origins of the Separate But Equal Doctrine*. New York: Da Capo Press, 1974.

Lofgren, Charles A. *The* Plessy *Case: A Legal-Historical Interpretation*. New York: Oxford University Press, 1987.

Mack, Raymond W., ed. *Our Children's Burden: Studies of Desegregation in Nine American Communities*. New York: Random House, 1968.

McNeil, Genna Rae. *Groundwork: Charles Hamilton Houston and the Struggle for Civil Rights*. Philadelphia: University of Pennsylvania Press, 1983.

Metcalf, George R. *From Little Rock to Boston: The History of School Desegregation*. Westport: Greenwood, 1983.

Miller, Arthur S. *Racial Discrimination and Private Education: A Legal Analysis*. Chapel Hill: University of North Carolina Press, 1957.

Miller, Loren. *The Petitioners: The Story of the Supreme Court and the Negro*. Cleveland: Meridian, 1967.

Myrdal, Gunnar. *An American Dilemma: The Negro Problem and Modern Democracy*. New York: Harper, 1944.

Namorato, Michael, ed. *Have We Overcome? Race Relations since* Brown. Jackson: University of Mississippi Press, 1979.

Nieman, Donald G. *Promises to Keep: African Americans and the Constitutional Order, 1776 to the Present*. New York: Oxford University Press, 1991.

Notre Dame Center for Civil Rights. *The Continuing Challenge: The Past and the Future of* Brown v. Board of Education: *A Symposium*. Evanston: Integrated Education Associates, 1975.

Orfield, Gary, and Susan E. Eaton, eds. *Dismantling Desegregation: The Quiet Reversal of* Brown v. Board of Education. New York: New Press, 1996.

Peltason, J. W. *Fifty-Eight Lonely Men: Southern Federal Judges and School Desegregation*. New York: Harcourt, Brace and World, 1961.

Pratt, Robert A. *The Color of Their Skin: Education and Race in Richmond, Virginia, 1954–89*. Charlottesville: University Press of Virginia, 1992.

Reams, Bernard D. Jr., and Paul E. Wilson, eds. *Segregation and the Fourteenth Amendment in the States: A Survey of State Segregation Laws, 1865–1953, Prepared for United States Supreme Court in re:* Brown v. Board of Education of Topeka. Buffalo: Hein, 1975.

Rosenberg, Gerald N. *The Hollow Hope: Can Courts Bring About Social Change*. Chicago: University of Chicago Press, 1991.

Salamone, Rosemary C. *Equal Education under Law: Legal Rights and Federal Policy in the Post-*Brown *Era*. New York: St. Martin's Press, 1986.

Sarat, Austin, ed. *Race, Law, and Culture: Reflections on* Brown v. Board of Education. New York: Oxford University Press, 1997.

"School Desegregation." *Howard Law Journal* 19 (Winter 1975): 1. Issue devoted to speeches presented at the 1974 NAACP National Education Conference held at Washburn University in Topeka, Kansas.

"School Desegregation: Lessons of the First Twenty-Five Years, Part 1," *Law and Contemporary Problems* 42, no. 3 (Summer 1978).

"School Desegregation: Lessons of the First Twenty-Five Years, Part 2," *Law and Contemporary Problems* 42, no. 4 (Autumn 1978).

Scott, Daryl Michael. *Contempt and Pity: Social Policy and the Image of the Damaged Black Psyche, 1880–1996.* Chapel Hill: University of North Carolina Press, 1997.

Smith, J. Clay Smith. *Emancipation: The Making of the Black Lawyer, 1844–1944.* Philadelphia: University of Pennsylvania Press, 1993.

Speer, Hugh W. *The Case of the Century: A Historical and Social Perspective on* Brown v. Board of Education of Topeka *with Present and Future Implications.* Kansas City: University of Missouri, Hugh W. Speer, 1968.

Stephan, Walter G., and Joe R. Feagin, eds. *School Desegregation: Past, Present, and Future.* New York: Plenum, 1980.

Sullivan, Patricia A. *Days of Hope: Race and Democracy in the New Deal Era.* Chapel Hill: University of North Carolina Press, 1996.

Tushnet, Mark. *Making Civil Rights Law: Thurgood Marshall and the Supreme Court.* New York: Oxford University Press, 1944.

———. *The NAACP's Legal Strategy against Segregated Education, 1925–1950.* Chapel Hill: University of North Carolina Press, 1987.

———, with Katya Lezin. "What Really Happened in *Brown v. Board of Education.*" *Columbia Law Review* 91, no. 8 (December 1991): 1867–1930.

Weinberg, Meyer. *Race and Place: A Legal History of the Neighborhood School.* Washington, D.C.: GPO, 1967.

Whitman, Mark. *Removing a Badge of Slavery: The Record of* Brown v. Board of Education. Princeton: Wiener Publications, 1992.

Wilkinson, J. Harvie III. *From* Brown *to* Bakke: *The Supreme Court and School Integration, 1958–1978.* New York: Oxford University Press, 1979.

Wollenberg, Charles. *All Deliberate Speed: Segregation and Exclusion in California Schools, 1855–1975.* Berkeley: University of California Press, 1976.

Woodward, C. Vann. *The Strange Career of Jim Crow.* 3rd ed. New York: Oxford University Press, 1974.

(Continued from p. iv)

Henry McNeal Turner, *Civil Rights: The Outrage of the Supreme Court of the United States Upon the Black Man* (1889). From *A Documentary History of the Negro People in the United States*, volume 1, Herbert Aptheker, editor. Copyright © 1961. Published by arrangement with Carol Publishing Group. A Citadel Press Book.

Justice Henry Brown's Decision. Reprinted with permission of the copyright holder, West Group, Cleveland Office.

Justice John Marshall Harlan's Dissent. Reprinted with permission of the copyright holder, West Group, Cleveland Office.

Letters of Negro Migrants of 1916–18. Courtesy of the Journal of Negro History, Morehouse College, Atlanta, Georgia.

Langston Hughes, "I, Too" (1926). From *Collected Poems* by Langston Hughes. Copyright © 1994 by the Estate of Langston Hughes. Reprinted with permission of Alfred A. Knopf, Inc.

W. E. B. Du Bois, "Does the Negro Need Separate Schools?" (1935). W. E. B. Du Bois, "Does the Negro Need Separate Schools?", *Journal of Negro Education* 4 (1935), 328–35. Copyright © 1935 by Howard University. All rights reserved.

Gunnar Myrdal, from *An American Dilemma: The Negro Problem and Modern Democracy* (1944). Excerpt submitted from *An American Dilemma: The Negro Problem and Modern Democracy by Gunnar Myrdal.* Copyright © 1944, 1962 by Harper & Row, Publishers, Inc. Reprinted by permission of HarperCollins Publishers, Inc.

Mississippi Voter Registration Form (1955). Manuscript Division, Library of Congress.

Judge John J. Parker's Decision. Reprinted with permission of the West Group.

Judge J. Waties Waring's Dissent. Reprinted with permission of the West Group.

Chief Justice Earl Warren's Decision (May 17, 1954). Leon Friedman, ed., *Argument: The Oral Argument Before the Supreme Court in Brown v. Board of Education of Topeka, 1952–1955* (New York: Chelsea House, 1969), pp. 325–31.

"All God's Chillun." From The New York *Times*, May 18, 1954. Copyright © 1954, 1964, 1974, 1979, 1984, 1994 by The New York Times Company. Reprinted by permission.

"The Decision of A Century." From The *Atlanta Daily World*, May 18, 1954. Atlanta Daily World.

"The Second Emancipation." From *The Amsterdam News*, May 22, 1954. Courtesy of *The Amsterdam News*.

"Will Stun Communities." From *The Pittsburgh Courier*, May 19, 1954. Reprinted by permission of GRM Associates, Inc., Agents for *The Pittsburgh Courier*.

"End of Dual Society." From the *Chicago Defender*, May 18, 1954. Reprinted with permission from the *Chicago Defender*.

"Emancipation." From *The Washington Post and Times Herald*, May 18, 1954. Copyright © *The Washington Post*.

"The Supreme Court Has Given Us Time." From the *Atlanta Constitution*, May 18, 1954. Reprinted with permission of the *Atlanta Constitution*.

"The Pattern of the Future." From the *Arkansas Gazette*, May 18, 1954. *Arkansas Democrat-Gazette*, 1997.

"Equality Redefined." From the *Boston Herald*, May 18, 1954. Reprinted with permission of the *Boston Herald*.

"The Segregation Decision." From the *Los Angeles Times*, May 19, 1954. Copyright, 1954, *Los Angeles Times*. Reprinted by permission.

" 'Violates' Way of Life." From *Cavalier Daily*, University of Virginia, May 18, 1954. Courtesy of the *Cavalier Daily*.

"Adjustment Held Difficult." From *Mississippian*, University of Mississippi, May 18, 1954. Courtesy of University of Mississippi Student Media Center.

Lillian Smith, The New York *Times*, May 31, 1954. Lillian E. Smith Estate.

Zora Neale Hurston, *Orlando Sentinel*, August 11, 1955. The *Orlando Sentinel*, August 11, 1955. Reprinted by permission.

San Francisco Chronicle, May 18, 1954. © *San Francisco Chronicle.* Reprinted with permission.

The Christian Science Monitor, May 22, 1954. © 1954 *The Christian Science Monitor.* Reprinted with permission.

The Arkansas Democrat, May 23, 1954. *Arkansas Democrat-Gazette,* 1997.

The Nashville Tennessean, May 23, 1954. Copyrighted by *The Tennessean,* May 23, 1954.

The Richmond Afro-American, May 22, 1954. Afro-American Newspapers Archives and Research Center.

Chicago Defender, June 12, 1954. Reprinted with permission from the *Chicago Defender.*

White Citizen's Council Notice. Courtesy of Professor William Moore, College of Charleston.

"Decade of Desegregation," New York *Times,* May 17, 1964; "Two Decades Later," New York *Times,* May 17, 1974; "An Age of Liberation," New York *Times,* May 17, 1979; "The Enduring Promise of 'Brown,' " New York *Times,* May 17, 1984; "Forty Years and Still Struggling," New York *Times,* May 18, 1994. Copyright © 1954, 1964, 1974, 1979, 1984, 1994 by The New York Times Company. Reprinted by permission.

Pictures

Removing segregation signs on Dallas buses. Corbis-Bettmann.

Separate entrance at a Florida theater and separate drinking fountains. Corbis-Bettmann.

Linda and Terry Brown. Carl Iwasaki, Life Magazine © Time Inc.

Class of a one-room school house. Courtesy of the Tuskegee Institute Archives.

FSA photograph of a black schoolhouse. Library of Congress. Arthur Rothstein, photographer. LC-USF34T01-25378-D.

Index

247

Printed in the United States
By Bookmasters